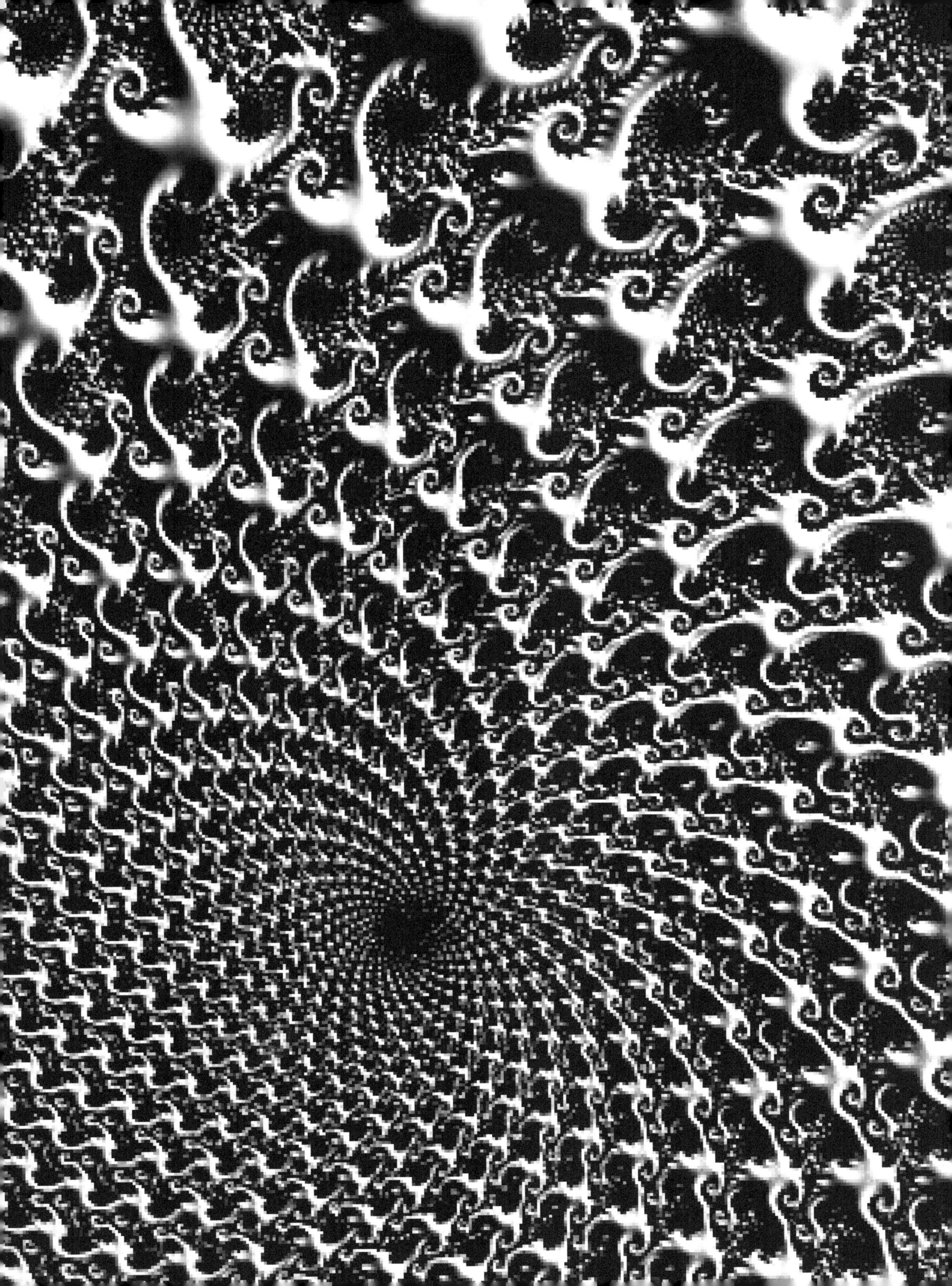

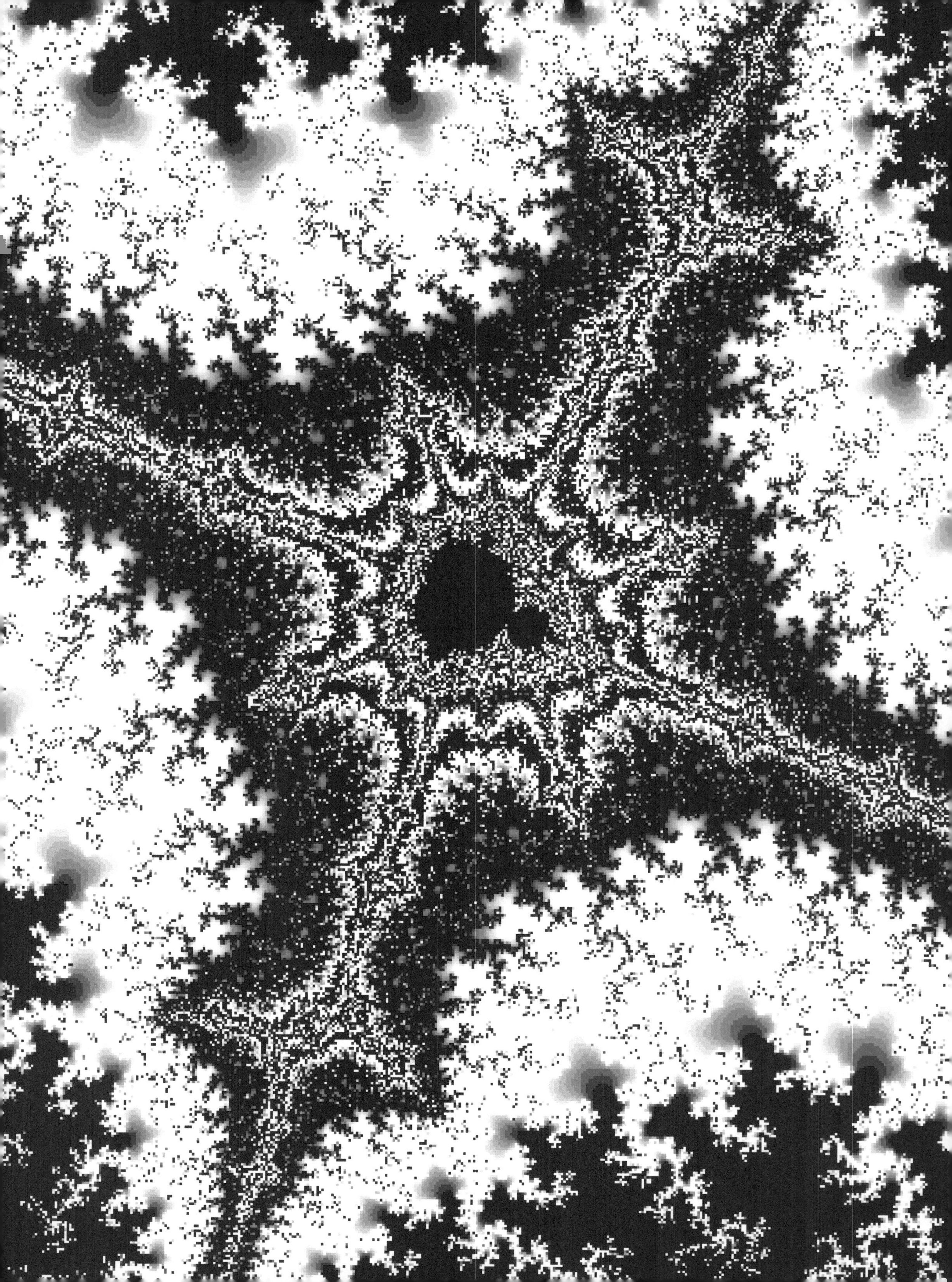

DOBSON'S ARC

Historical Roots of STEM
from Antiquity to the Present Day

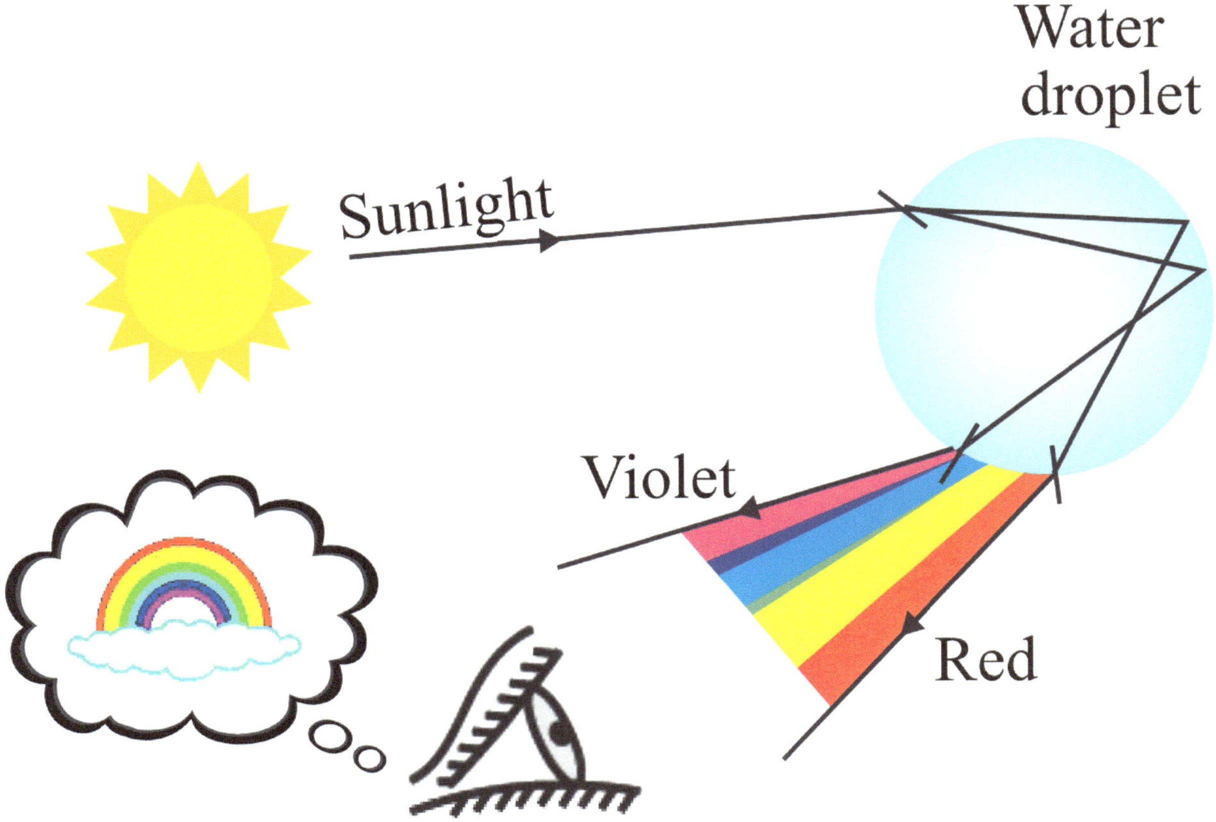

Mixed Media Works
by Duncan Dobson

Edited by Lucy Dobson

DOBSON'S ARC

Mixed Media Works by Duncan Dobson
Edited by Lucy Dobson

Publisher: Duncan Dobson

American Poetry Systems
South San Francisco
California
USA
www.apsssf.com
www.Etsy.com/shop/AmPoSys
APS#006
First Edition 2024
Second Edition 2025
10 8 6 4 2

Other books from American Poetry Systems:

APS#001 *Second Sum* by Duncan Dobson
APS#002 *Poems* by Daniel Farnan
APS#003 *The Two Wolves* by Anya March
APS#004 *Fried Skin* by Duncan Dobson
APS#005 *48 Letters* by Duncan Dobson

Submit your poetry or other work for publication!
For up-to-date information
on which original paintings
are still available, please visit
www.Etsy.com/shop/NewUnitedMedia.

Printed with ease at Lulu.com.
ISBN 979-8-218-48223-7

Dedicated to my father, Duncan C. Dobson, Jr., my junior high math teacher Mr. Dolyniuk and my high school math teacher Mr. Kunkle.

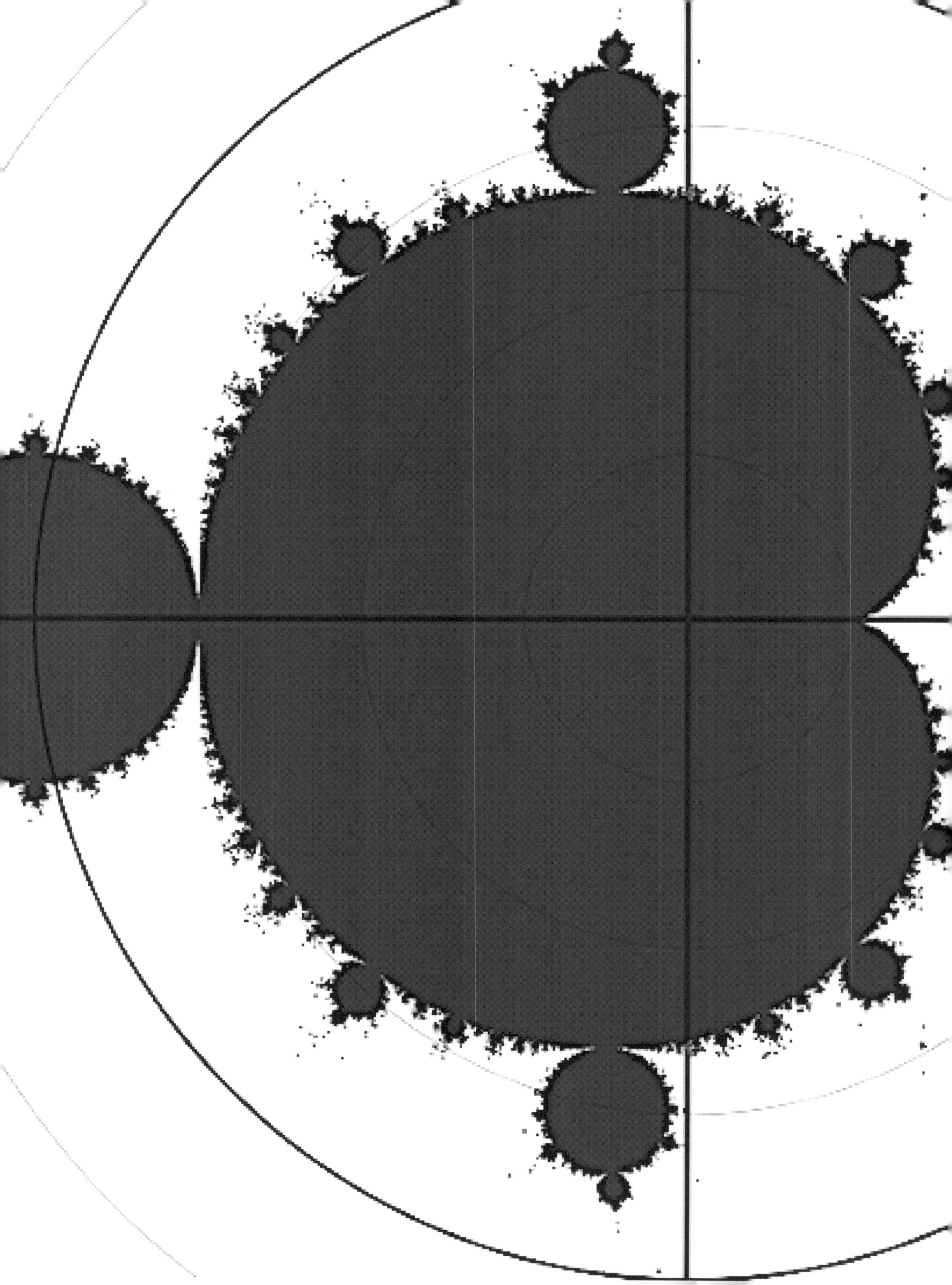

Table of Contents

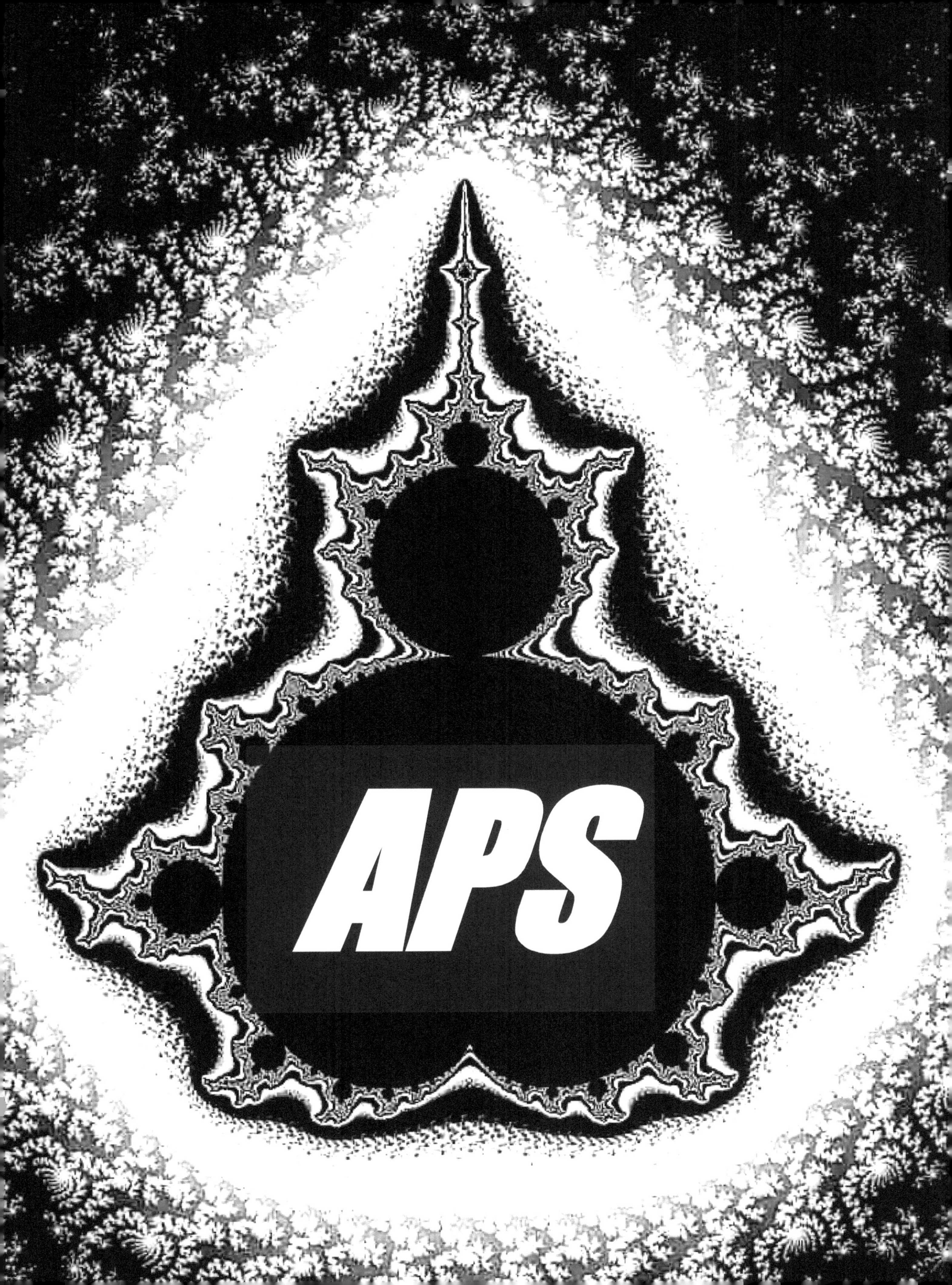
APS

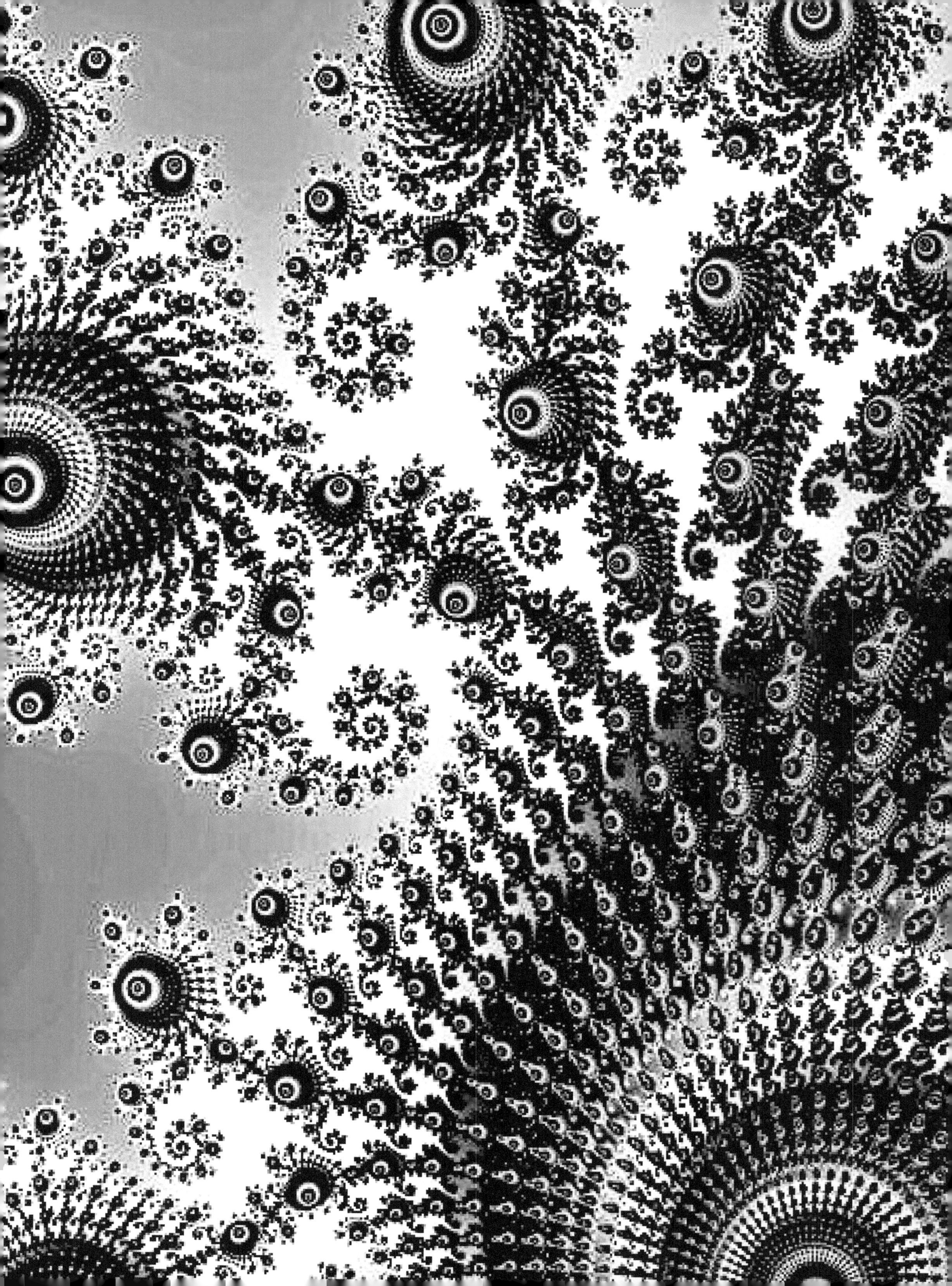

PREFACE

This volume presents a chronological list of mathematicians, physicists, and astronomers from antiquity to the present day. Their different ways of appraising the world are all ultimately tied to the same thing: understanding the physical universe that confronts all of us, every day, all our waking lives. If you don't think it's important to investigate the fundamental principles of the physical universe that we all inhabit, consider that the physical world has a great deal to say about whether you (or any of us) live or die at each and every moment, in every instance, all the time. Please be assured that as far as any Earthly astronomer knows, a huge asteroid from outer space could hit the Earth at any time and wipe us all out right now. There may be a massive earthquake at any moment. There may be a devastating hurricane coming. A blood vessel might burst in your sibling's brain. It is the job of science to understand such natural phenomena. At its best, science is a desperate struggle.

Personally, I do think there is more going on in the world than the physical sciences can explain. Human consciousness does not make sense as a merely natural phenomenon. If we were all just atoms bouncing around according to natural physical laws, how then could those atoms ever overcome, control, and direct those very laws? Yet this is precisely what humans routinely do by exerting our own individual wills to the extent that we can. Human nature itself defies pure materialism. There must be a literally supernatural intelligence, an intelligence that precedes and is above human intelligence. Human intelligence must have an intelligent creator.

As far as contact from alien life outside the Earth, I haven't seen any convincing sign of it. I think most people fail to appreciate just how far one star is from another. With our current technology, even if we could muster the effort, it would take us tens of thousands of years to get to the nearest star to our sun, that is if we could survive the lack of gravity and intense radiation of interstellar space. And that's just the closest star! If you think there will be a warp drive or teleportation or wormholes or stargates or some other kind of instantaneous form of travel between solar systems anytime soon, you are comically deluded. Why don't you watch another Star Wars movie?

AUTHOR'S NOTE

I developed this style of portraiture over the course of 2017 and 2018 while living in San Francisco, CA. A good deal of trial and error was involved. During the Covid-19 pandemic of 2020 and 2021, I made great strides in the technique. Now, the style is basically a formula that can be applied to any personage. At first, I mostly chose to portray men of science because I idolized them in my youth. This is due to my father, who would fall asleep at night reading Richard Feynman lectures. He also taught me some fundamentals of calculus at an early age. I have since branched out and done portraits of some of my favorite philosophers, authors and musical artists.

I do not have any degrees in the sciences. I am neither a mathematician nor a physicist. A lot of what is in this book is taken from the internet without permission, but I would plead fair educational use. I have mined the internet for content and incorporated it into a painterly context. I am an artist, not a scientist, but I feel I have a good grasp of the Western scientific tradition.

This book should not be thought of as an exhaustive list. There are any number of other historical figures that could be included. Maybe I'll get around to making pictures of them someday. When I started making these portraits, I did not imagine that I would collect them into a book. The selections I have made here undoubtedly reflect my own biases. So perhaps it is an idiosyncratic list, but I still feel it represents the tradition pretty well.

AN APOLOGY

One thing that is obviously omitted from this book is any reference to the scientific achievements of the ancient Chinese. While such achievements may exist, they did not really influence the Indo-European tradition that I am depicting here. The ancient Chinese seem to have invented the metal plow, gunpowder, and many other early technologies. Even today, however, its obscurity and a formidable language barrier prevent many in the West from understanding ancient Chinese mathematics in any detail. I think Chinese science was more about alchemy, medicine, numerology, and astrology than mathematics, per se. There is much to be learned there, but it's not relevant to the topic at hand. Apologies to the aggrieved.

DOBSON'S ARC

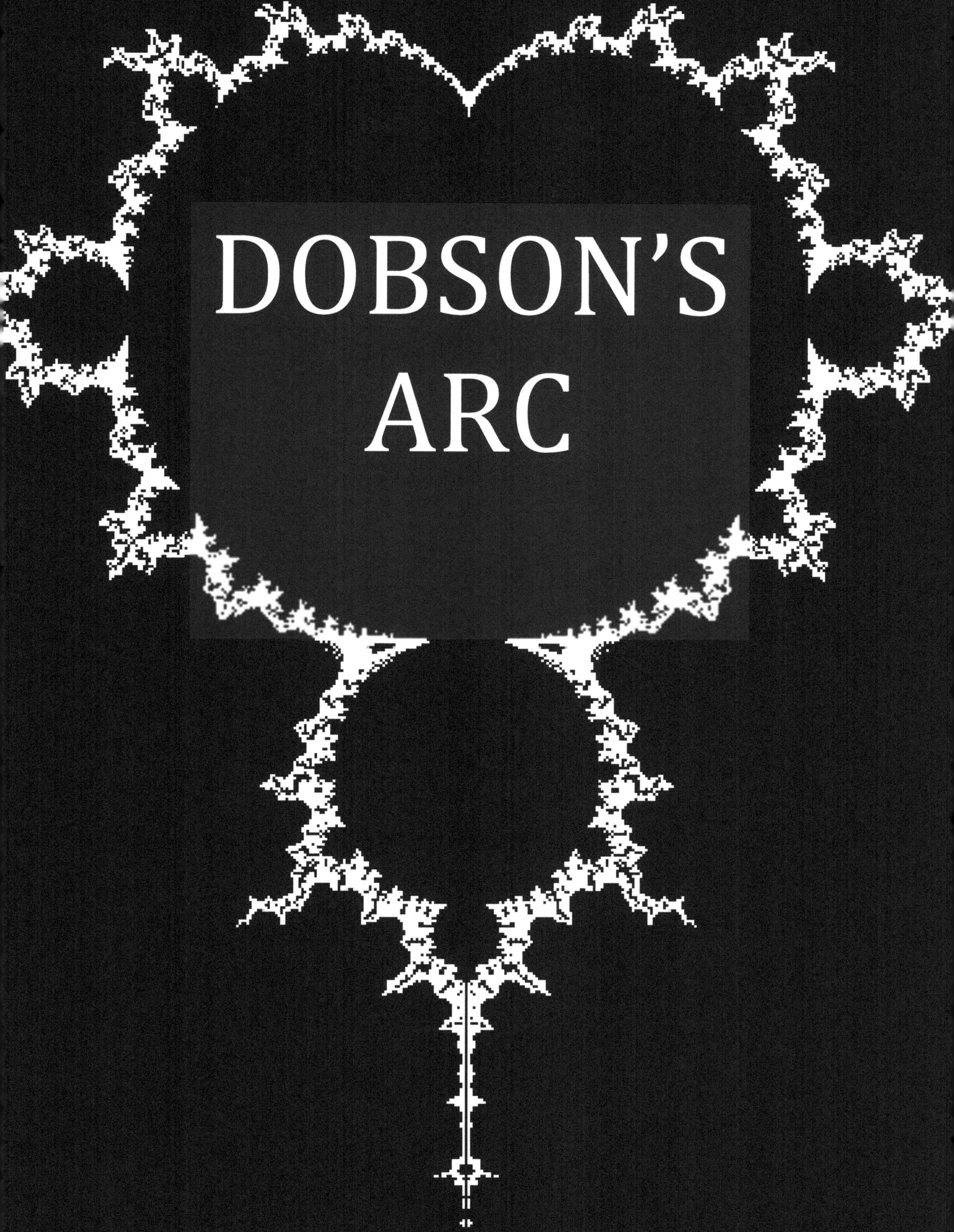

Pythagoras (c. 500 BCE)
Greek

Pythagoras was the semi-mythical Ancient Greek founder of Western mathematics. He was not the first to invent his theorem (both the Egyptians and the Babylonians likely knew it before him), but it bears his name in all the Western world. Pythagoras' cult revered number as represented by music, and by string lengths and the ratios between them. Pythagoras thought that irrational numbers such as π and the square root of two were heresy. His cult was also vegetarian and involved a five-year vow of silence for initiates. It is said that Pythagoras met his end when he refused to cross a beanfield while being chased by a ravaging band of democrats, beans being a symbol of democracy. None of Pythagoras' written work survives, and it is not known if he produced any.

The Pythagorean Theorem

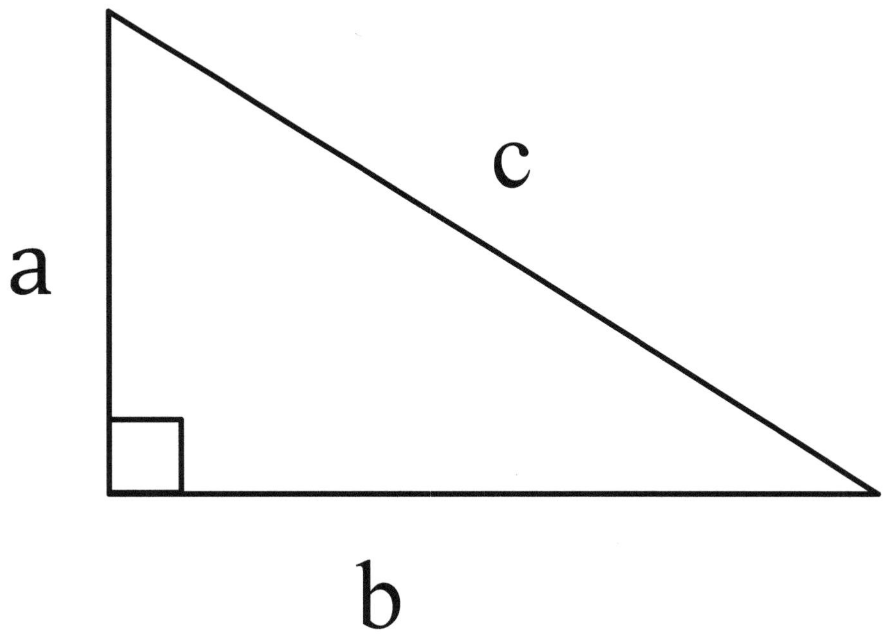

$$a^2 + b^2 = c^2$$

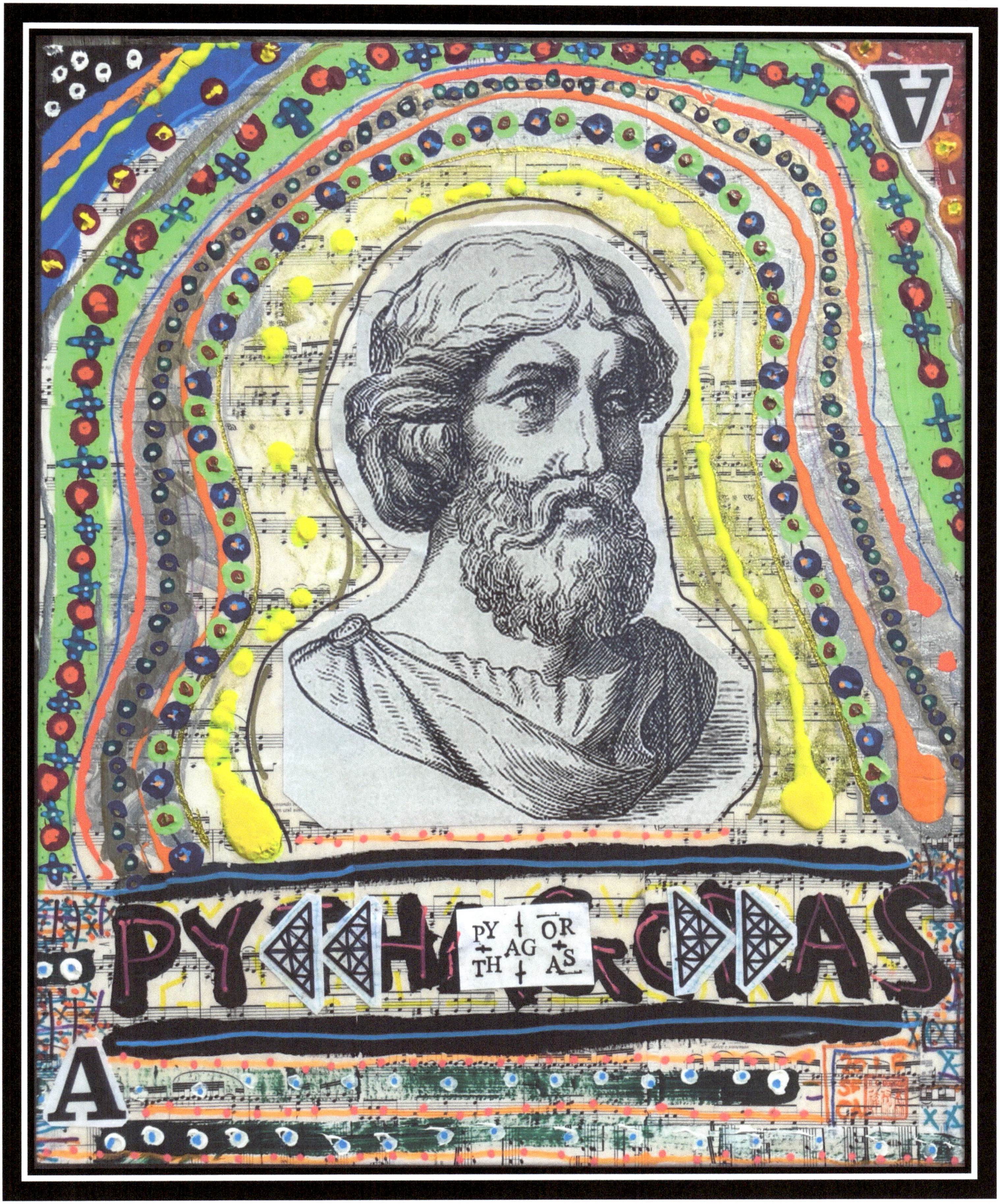
A
PY
OR
AG
TH
AS
A

Euclid (c. 300 BCE)

Greek

Euclid remains and always will be the Father of Geometry, and so he is the true father of the Western mathematical tradition. His book *Elements* was the standard of mathematics for 2000 years.

Euclid's Five Postulates of Geometry are, in order:

1. A straight line segment may be drawn from any given point to any other.

2. A straight line may be extended to any finite length.

3. A circle may be described with any given point as its center and any distance as its radius.

4. All right angles are congruent.

5. If a straight line intersects two other straight lines, and so makes the two interior angles on one side of it together less than two right angles, then the other straight lines will meet at a point if extended far enough on the side on which the angles are less than two right angles.

None of these postulates can be argued, except the fifth, the so-called Parallel Postulate, which became a bone of contention in mathematics. Eventually, around 1850, it was shown that there are non-Euclidean geometries such as spherical and hyperbolic geometries. In such curved geometries, the parallel lines of the flat Euclidean plane can intersect.

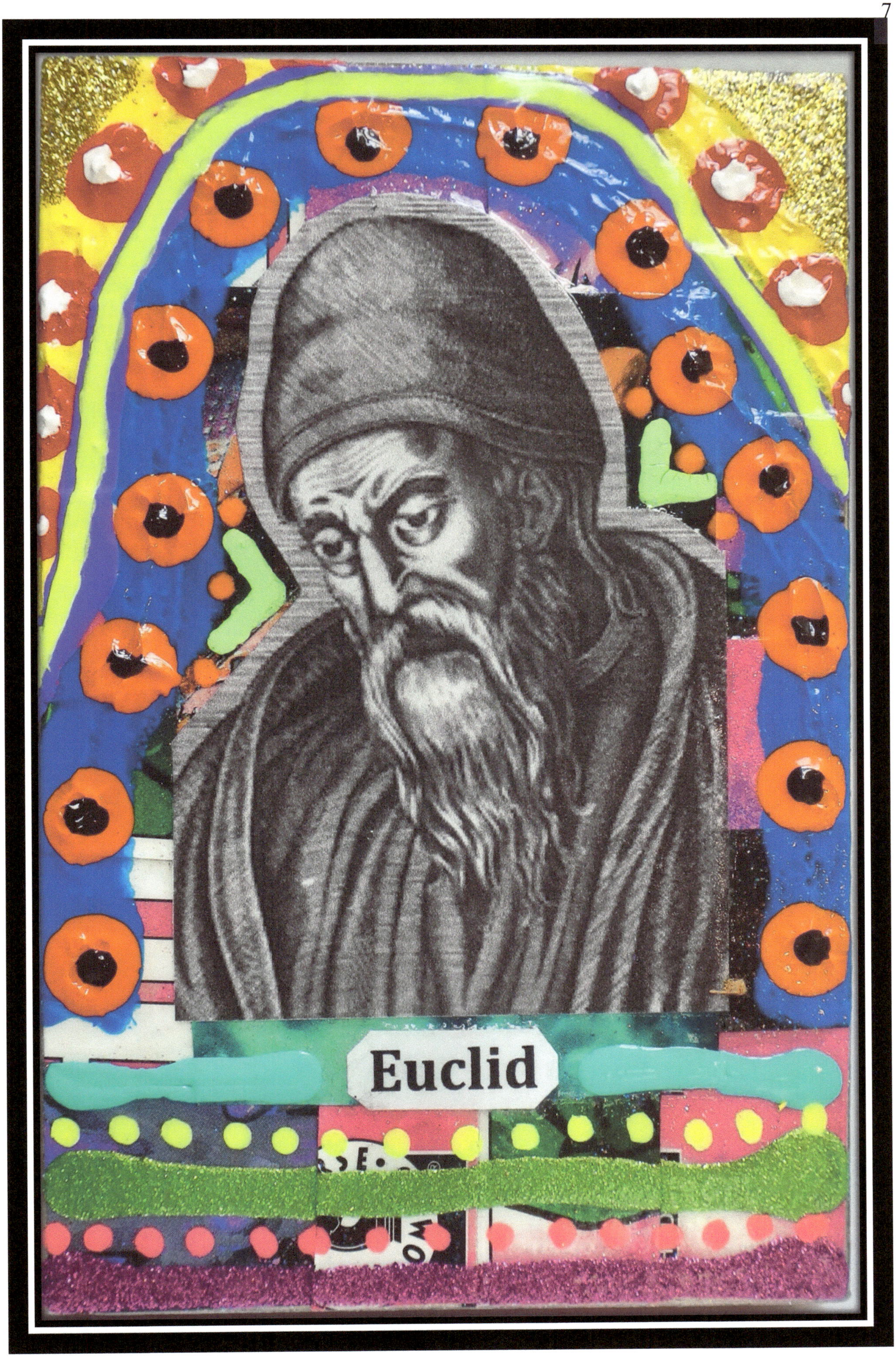
Euclid

Archimedes (287-212 BCE)

Greek

Archimedes was an Ancient Greek mathematician, physicist, engineer, astronomer, and inventor from the city of Syracuse in Sicily. A famous anecdote about Archimedes is that the King asked him to determine if his royal crown was pure gold or if it had been adulterated with other metals. Since the crown was an intricate, irregular shape, the problem was quite puzzling. Then, one day, Archimedes drew a bath. As he sank into the water, he realized the answer to the problem! Forgetting to dress himself, he ran into the street naked shouting, "Eureka!" or "I have found it!" He had realized that the amount of displaced water was equal to the volume of the body submerged. This would allow him to measure the volume of the crown and, knowing its weight, find its density and thus its metallic composition.

"Give me a lever long enough and a fulcrum on which to place it, and I shall move the Earth!"

-Archimedes

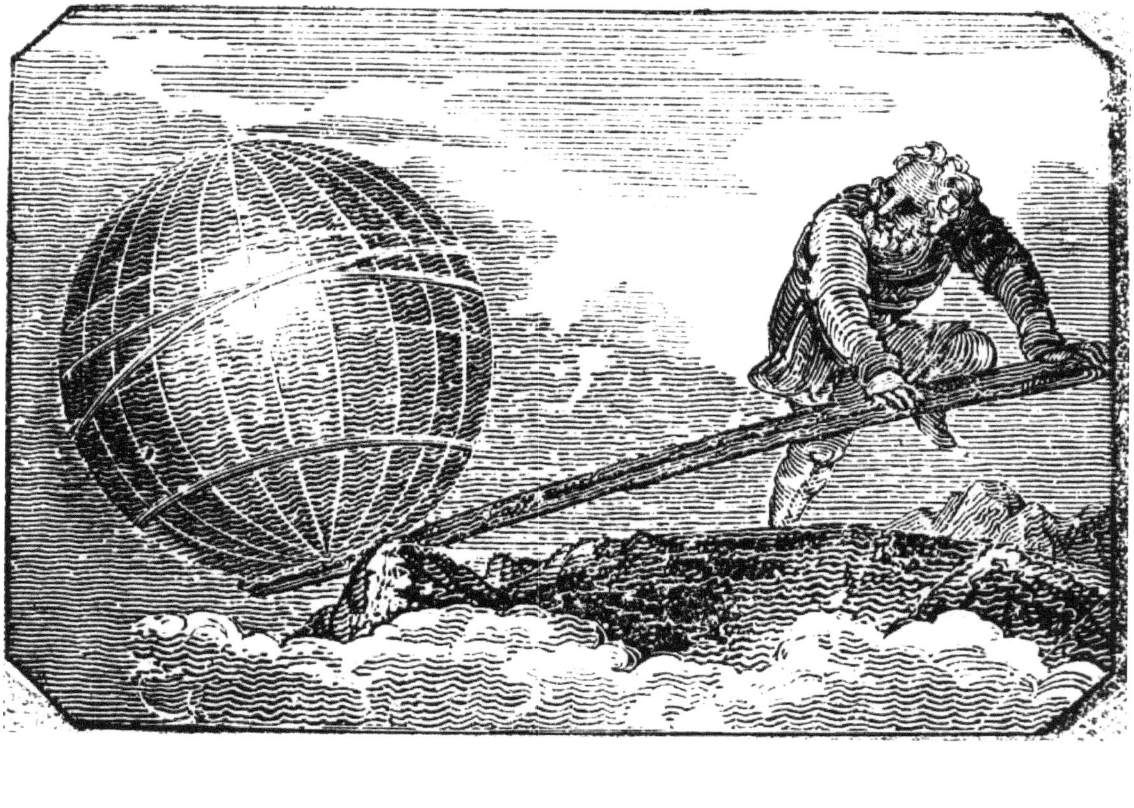

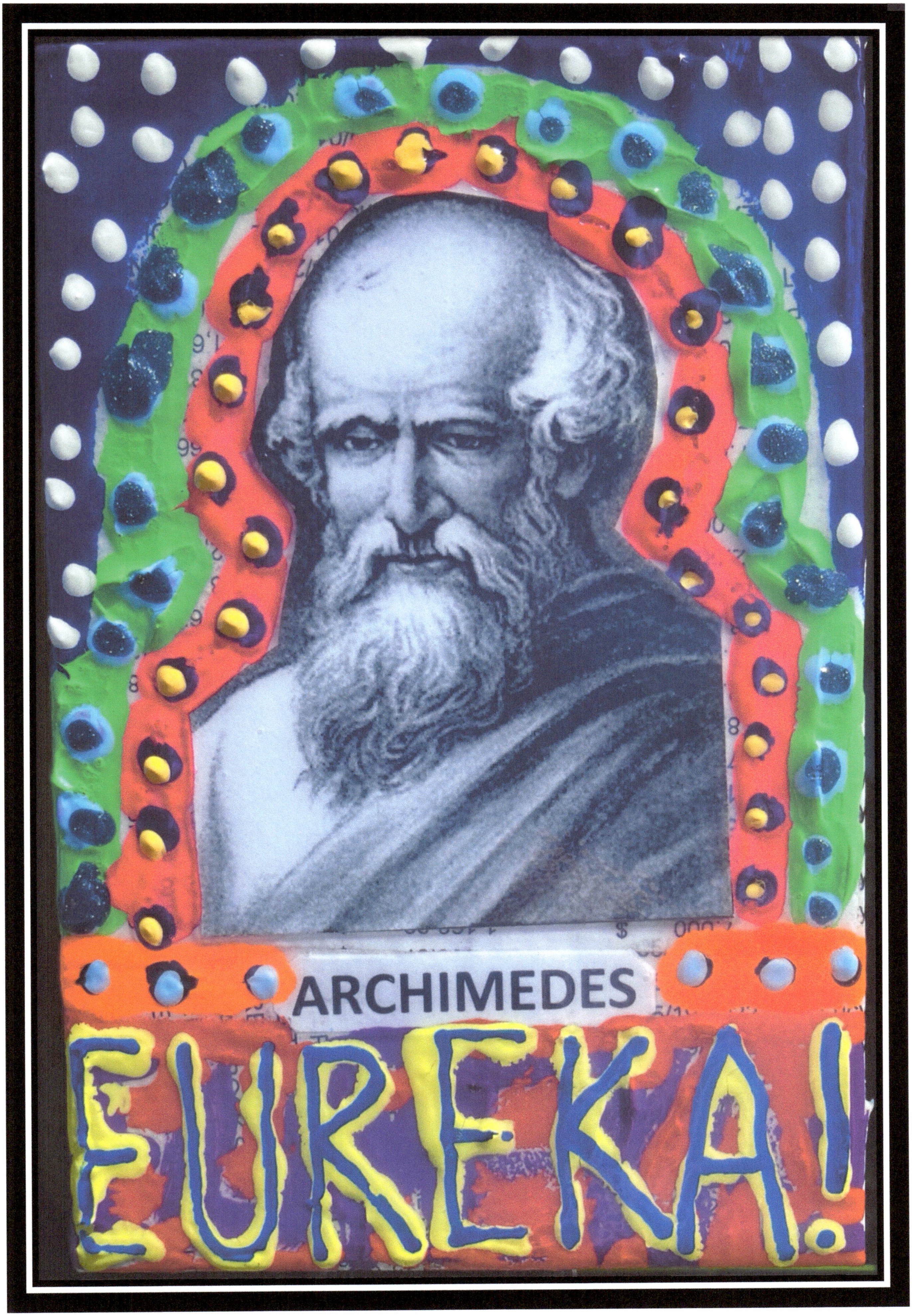
ARCHIMEDES
EUREKA!

Aryabhata (476-550) and Brahmagupta (598-668)
Indian

Two Hindu thinkers in the golden age of Indian science. Zero. Negative numbers. Trigonometry. Numerals. Eclipses.
I really couldn't say it better myself.
"Aryabhata discussed the concept of sine in his work by the name of *ardha-jya*, which literally means 'half-chord'. For simplicity, people started calling it *jya*. When Arabic writers translated his works from Sanskrit into Arabic, they referred it as *jiba*. However, in Arabic writings, vowels are omitted, and it was abbreviated as *jb*. Later writers substituted it with *jaib*, meaning 'pocket' or 'fold (in a garment)'. (In Arabic, *jiba* is a meaningless word.) Later in the 12th century, when Gherardo of Cremona translated these writings from Arabic into Latin, he replaced the Arabic *jaib* with its Latin counterpart, *sinus*, which means 'cove' or 'bay'; thence comes the English word 'sine'." Wikipedia, "Aryabhata", Feb. 20, 2024.

Integers and numerals

-1/O/1

Eclipses

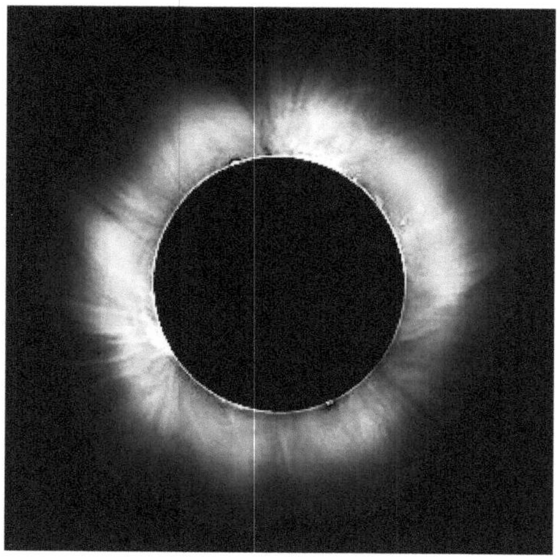

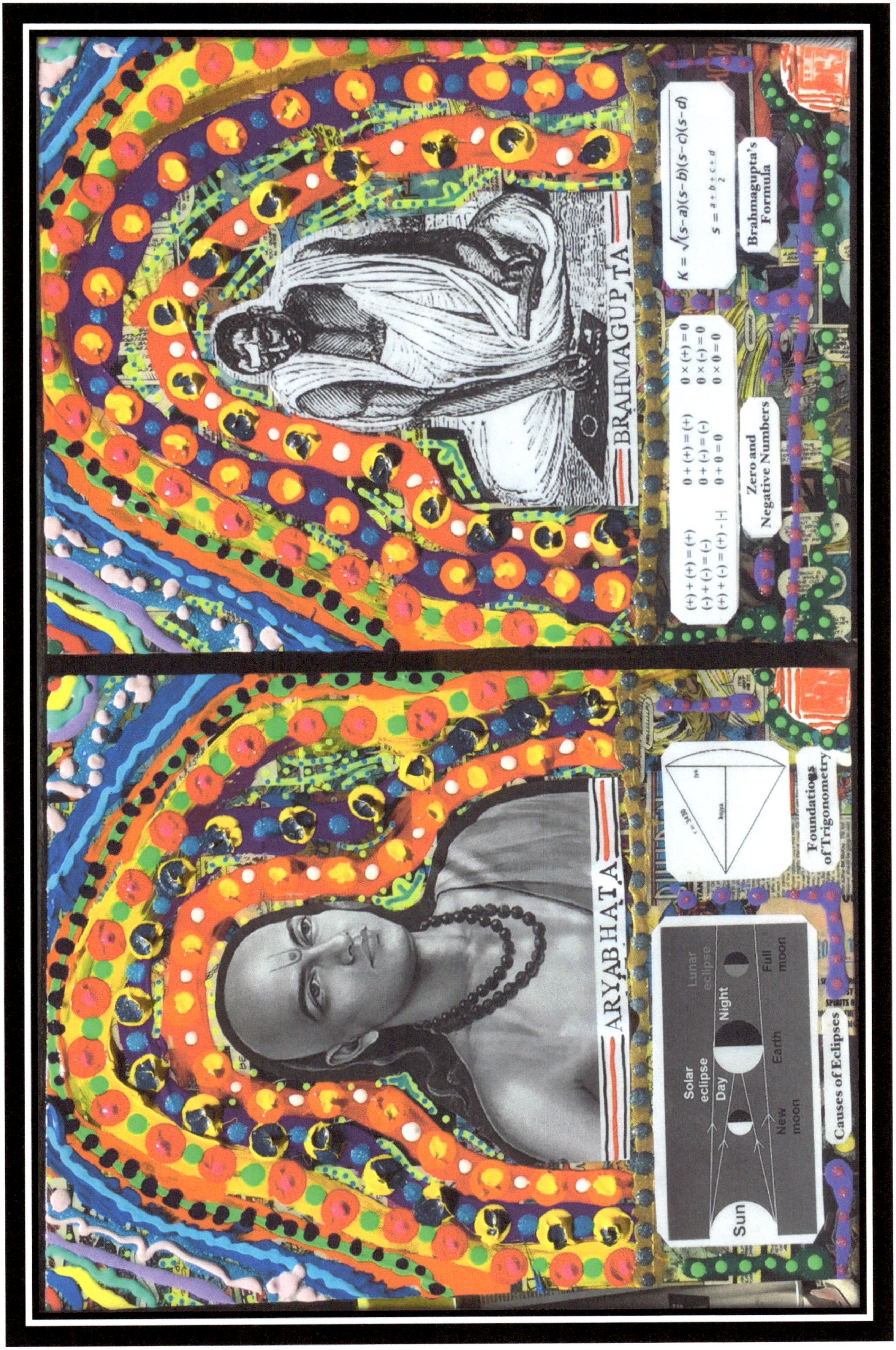
ARYABHATA
Causes of Eclipses
Sun
Solar eclipse
Day
Night
Lunar eclipse
New moon
Earth
Full moon
Foundations of Trigonometry
BRAHMAGUPTA
(+) + (+) = (+)
(-) + (-) = (-)
(+) + (-) = (+) - |-|
0 + (+) = (+)
0 + (-) = (-)
0 + 0 = 0
0 × (+) = 0
0 × (-) = 0
0 × 0 = 0
Zero and Negative Numbers
K = √(s−a)(s−b)(s−c)(s−d)
s = (a+b+c+d)/2
Brahmagupta's Formula

Omar Khayyam (1048-1131) and Al Khwarizmi (c. 780-850)

Persian

Two medieval Muslim polymaths. We get our word "algorithm" from Al Khwarizmi's name. He also gave us the word "algebra", which is derived from "al jabr", i.e., the reckoning, the balancing.
Omar Khayyam is also a famous poet. The early Muslims had the knowledge of both the ancient Indians and Greeks, which allowed them to innovate prolifically, not to mention the renewed emphasis on learning and knowledge that Islam had brought to their society. They also preserved the ancient Greek texts, allowing them to be reintroduced into Europe in the late Middle Ages.

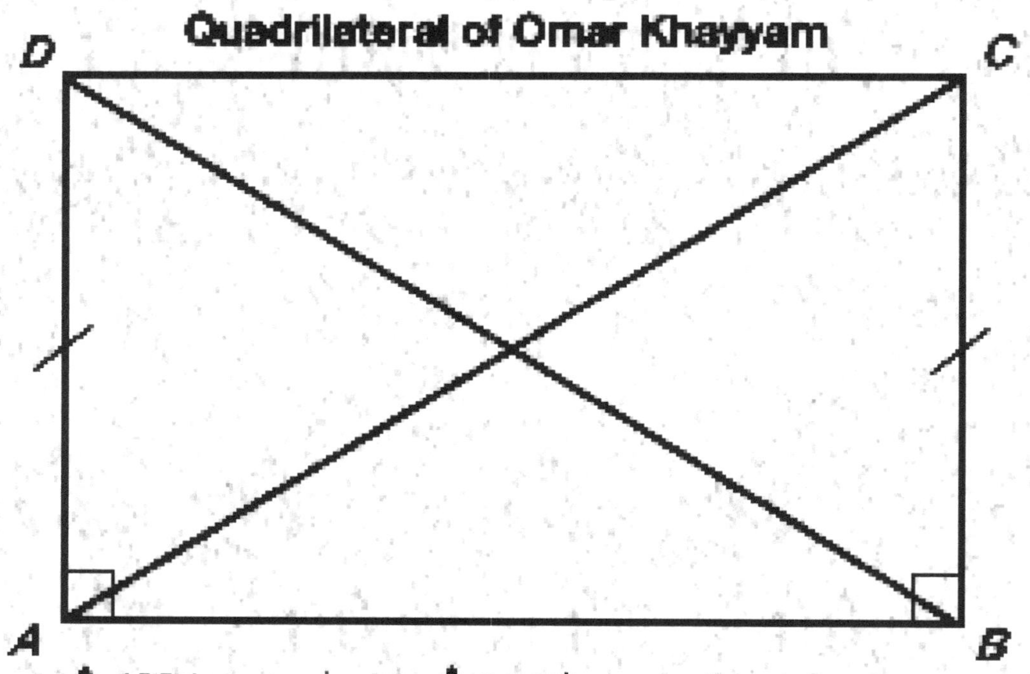

$\triangle ABD$ is congruent to $\triangle BAC$ (because they have two sides and the included angle that are equal, respectively). Hence, $AC = BD$, so $\triangle ADC$ is congruent to $\triangle BCD$ (the two triangles having three equal sides). Therefore, $\angle ADC = \angle BCD$.

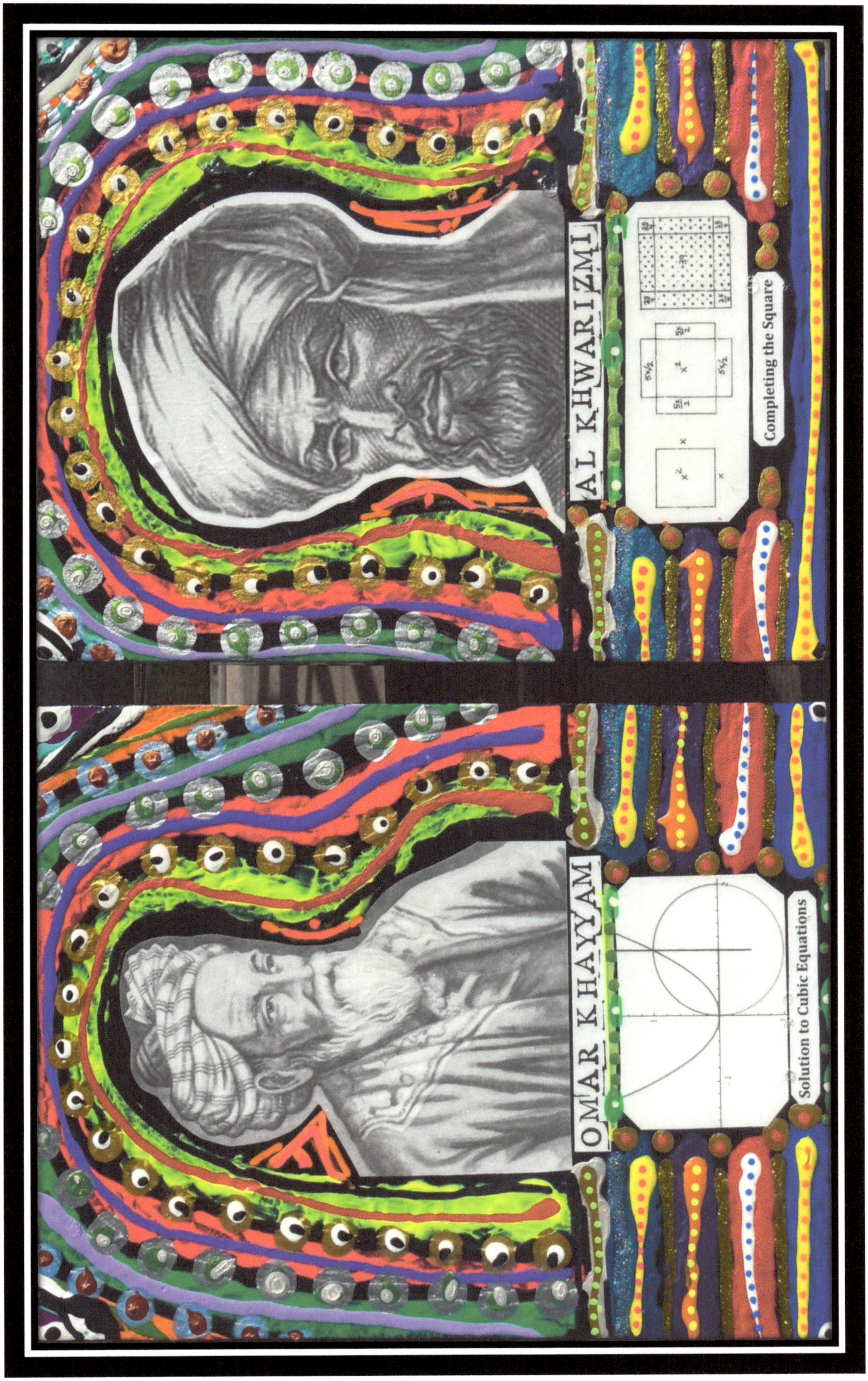
AL KHWARIZMI
Completing the Square
OMAR KHAYYAM
Solution to Cubic Equations

Fibonacci
aka Leonardo Bonacci, Leonardo of Pisa,
or Leonardo Bigollo Pisano (Leonardo the Traveler from Pisa)
(c. 1170-1250)
Italian

The Fibonacci Sequence. The Golden Ratio. European numerals. Fibonacci is symbolic of the introduction of ancient knowledge into Europe in the 12th and 13th centuries. This process laid the groundwork for the Renaissance, the Scientific Revolution, and the Enlightenment.

The Fibonacci Sequence

The Fibonacci Sequence is the list of numbers, starting from 0 or 1, where the next number is the sum of the previous two.

$$F_n = F_{n-1} + F_{n-2}$$

The Limit of the Fibonacci Sequence

As *n* gets very large, the ratio between one term and the next approaches ϕ, phi (fee), the Golden Ratio, an irrational number. The decimal approximation of ϕ is 1.6180339...

$$\lim_{n \to \infty} \frac{F(n+1)}{F(n)} = \phi$$

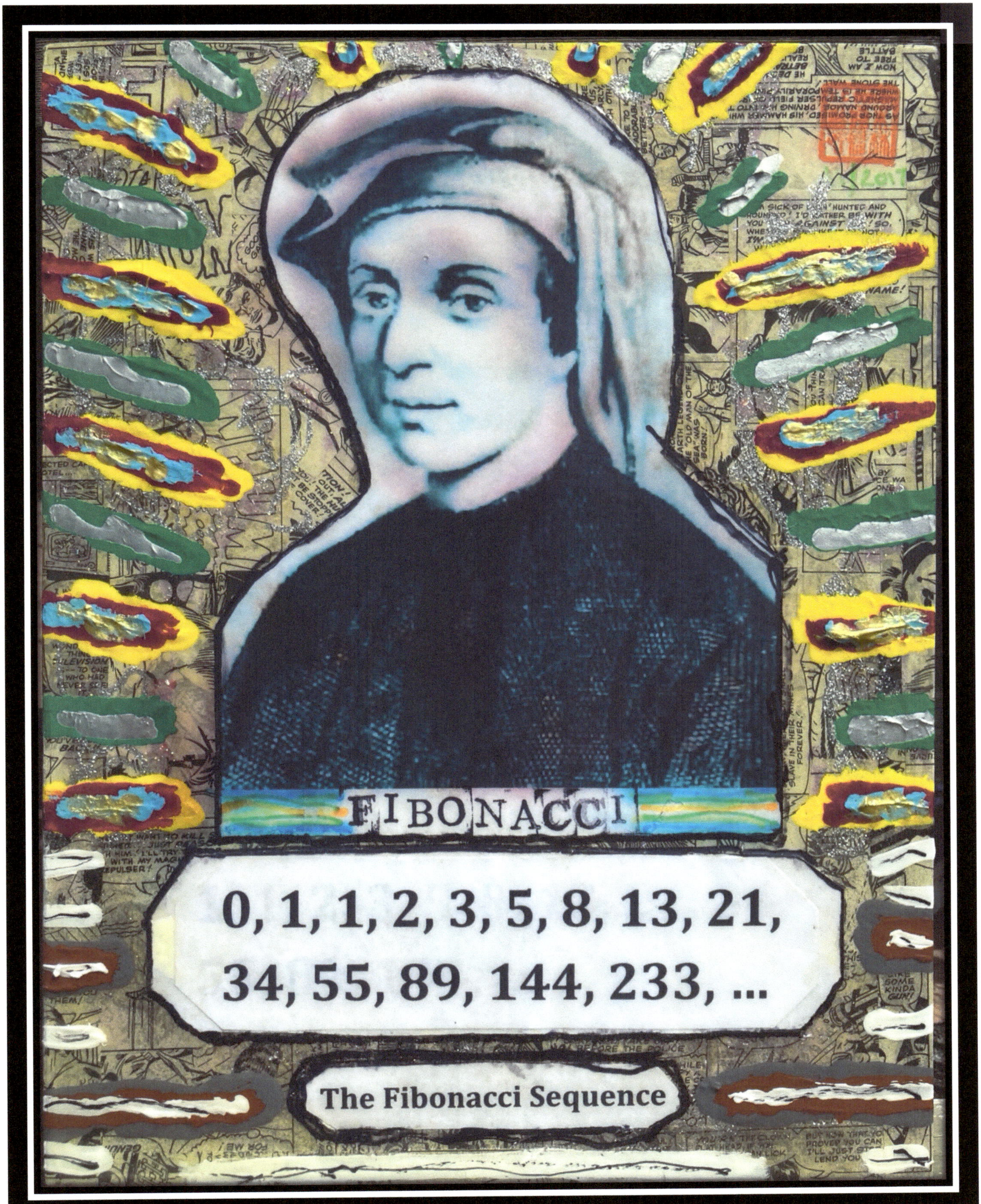
FIBONACCI
0, 1, 1, 2, 3, 5, 8, 13, 21,
34, 55, 89, 144, 233, ...
The Fibonacci Sequence

Johannes Kepler (1571-1630) & Galileo Galilei (1564-1642)
German (Kepler) and Italian (Galilei)

Kepler described the elliptical nature of planetary orbits. Building on the work of Copernicus before him, Kepler refined the theory of heliocentrism, the idea that the sun is at the center of our solar system, into one that accorded with observation.
Using new, state of the art telescopes, Galileo Galilei correctly surmised that the planet Jupiter has its own moons. This observation put an end to geocentrism in general, as it proved that not all heavenly bodies orbit the Earth.

Kepler's Solar System

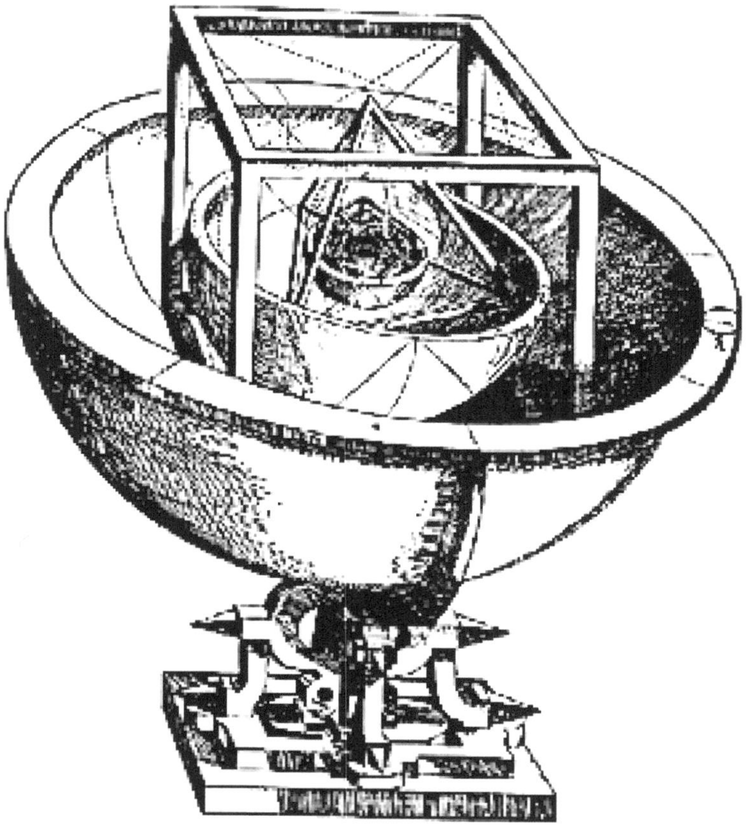

"Dropping science like Galileo dropped the orange!"

-The Beastie Boys

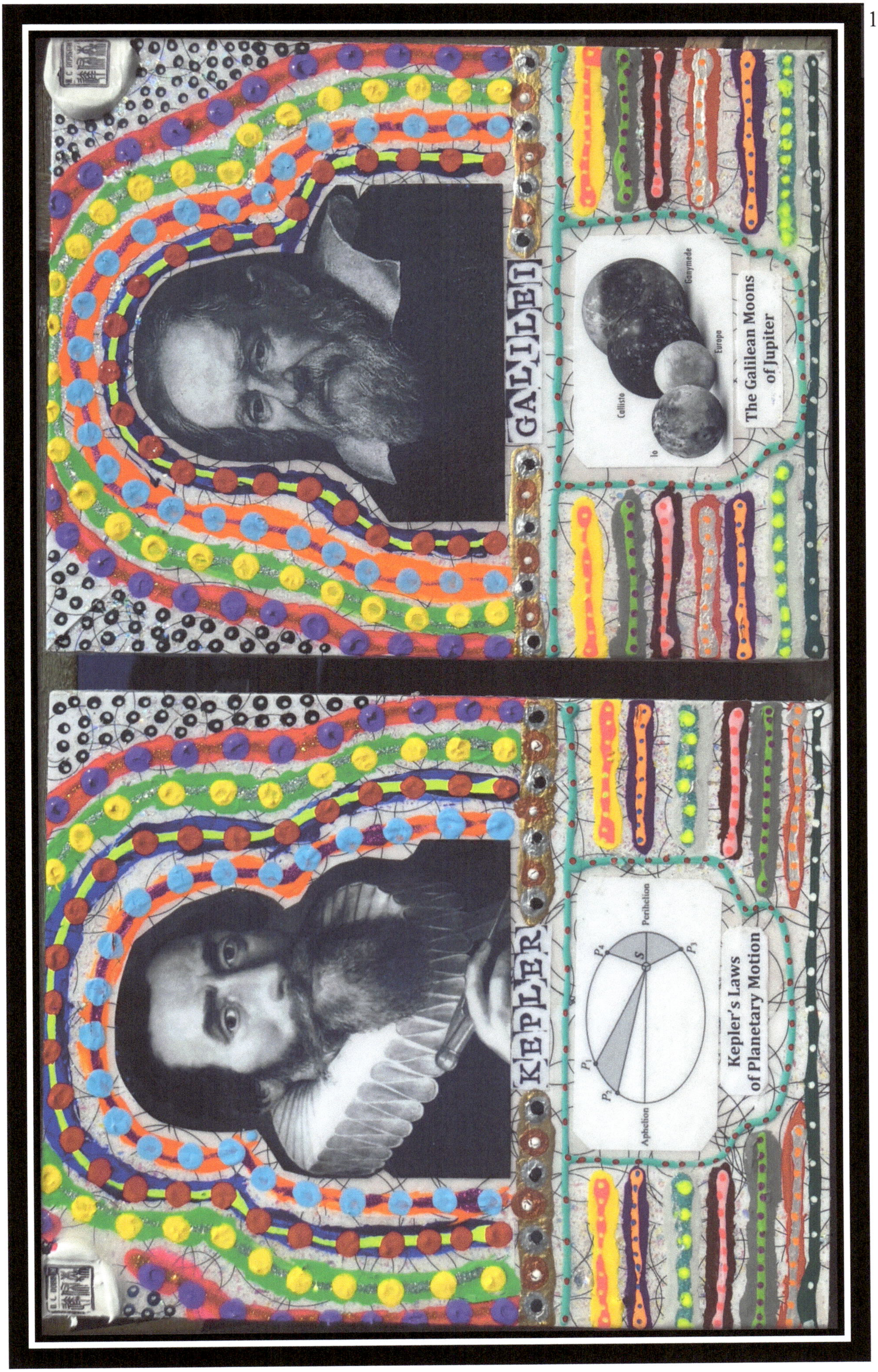
GALILEI
Callisto
Io
Europe
Ganymede
The Galilean Moons of Jupiter
KEPLER
Aphelion
Perihelion
Kepler's Laws of Planetary Motion

René Descartes (1596-1650) & Pierre de Fermat (1607-1665)
French

Descartes was a professional philosopher who used mathematics in his work. He traveled throughout his life in the reaches of Northern Europe. Descartes is often said to have been a great skeptic, but he was actually a committed Roman Catholic.
Fermat was a lawyer and a judge who also studied mathematics. Both were instrumental in laying the groundwork for calculus.

"Cogito, ergo sum."

(I think, therefore I am.)

-Descartes

Fermat's Principle

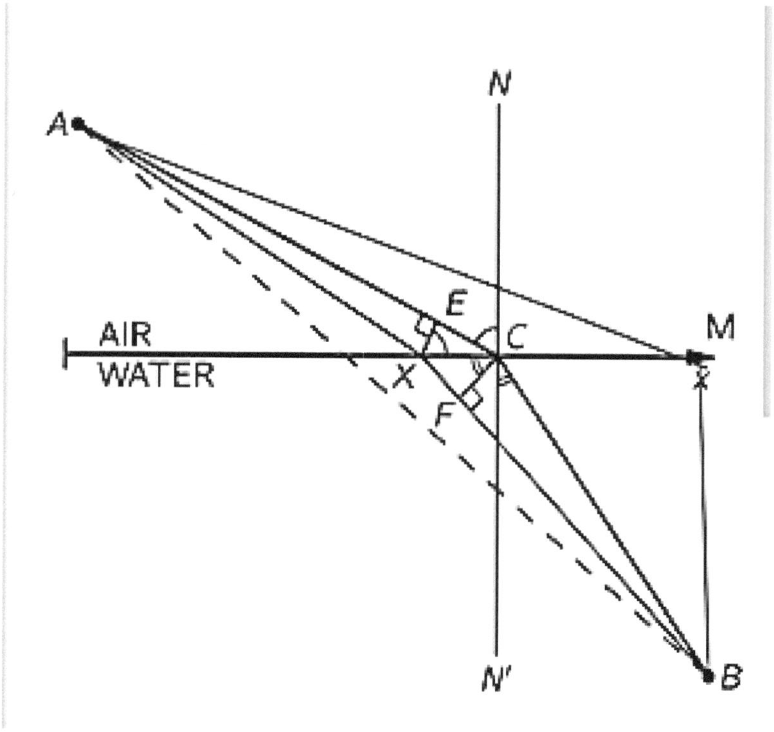

This principle has to do with the passing of light from air into water.

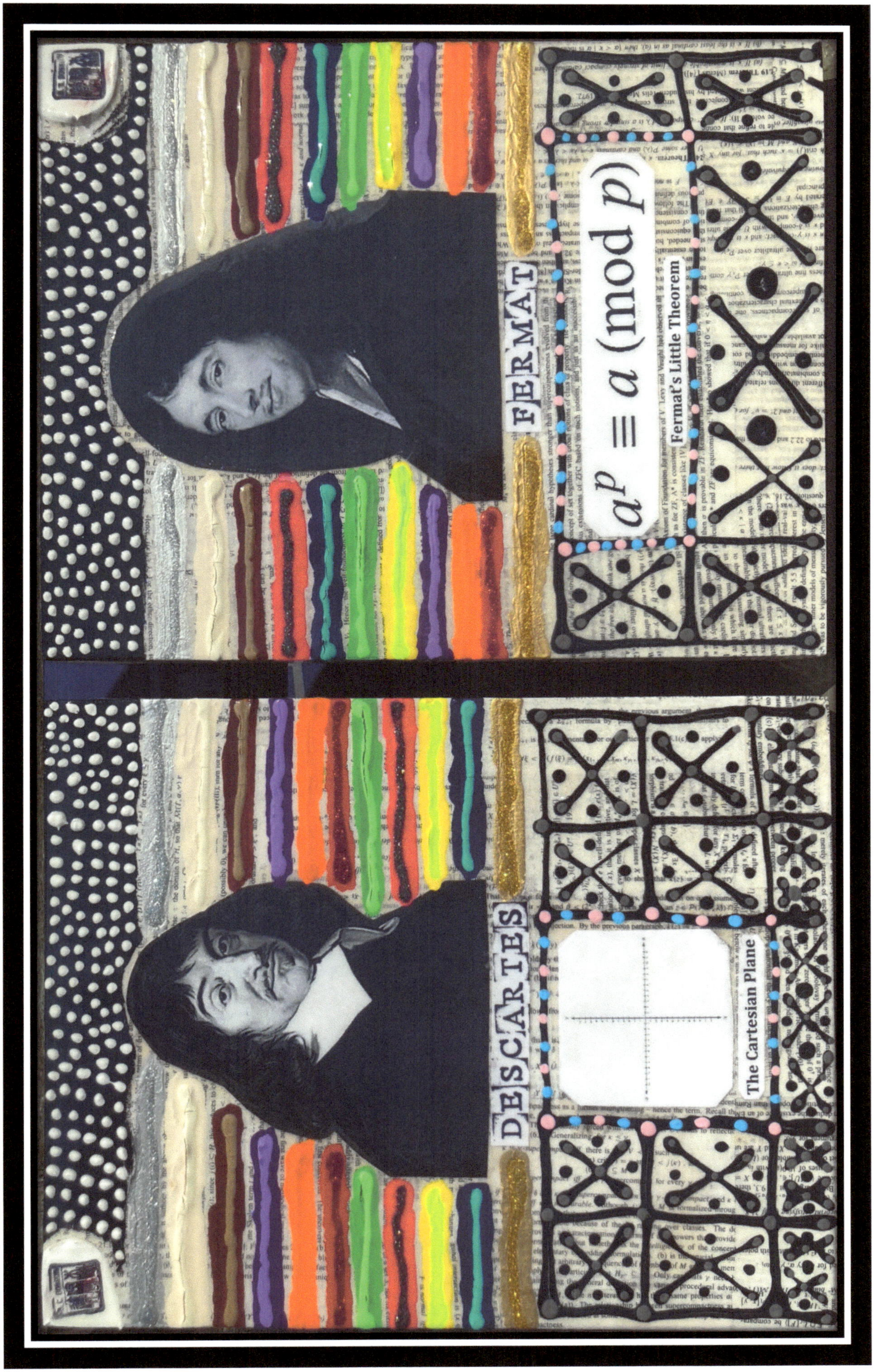

FERMAT
$a^p \equiv a \pmod{p}$
Fermat's Little Theorem
DESCARTES
The Cartesian Plane

Isaac Newton (1643-1727) and
Gottfried Wilhelm Leibniz (1646-1716)
English (Newton) and German (Leibniz)

Newton and Leibniz invented calculus independently at almost exactly the same time. Newton gets more credit in the English-speaking world, but to this day, mathematicians use Leibniz's notation. What is calculus? Calculus is the mathematics of infinitesimally small intervals. When we find a derivative, we are finding the slope of a curve at a vanishingly small point. When we take an integral, we are reducing slices of an area to zero width.

The Derivative of a Function

The slope of a curve at a point

1. Choose an interval
2. Find the raw change

$$f'(x) = \lim_{dx \to 0} \frac{f(x+dx) - f(x)}{dx}$$

4. Make your model perfect
3. Find the rate of change

The Integral of a Function

The area under the curve

$$\int_a^b f(x)dx = \lim_{n \to \infty} \sum_{i=1}^{n} \Delta x \cdot f(x_i)$$

where $\Delta x = \frac{b-a}{n}$ and $x_i = a + \Delta x \cdot i$.

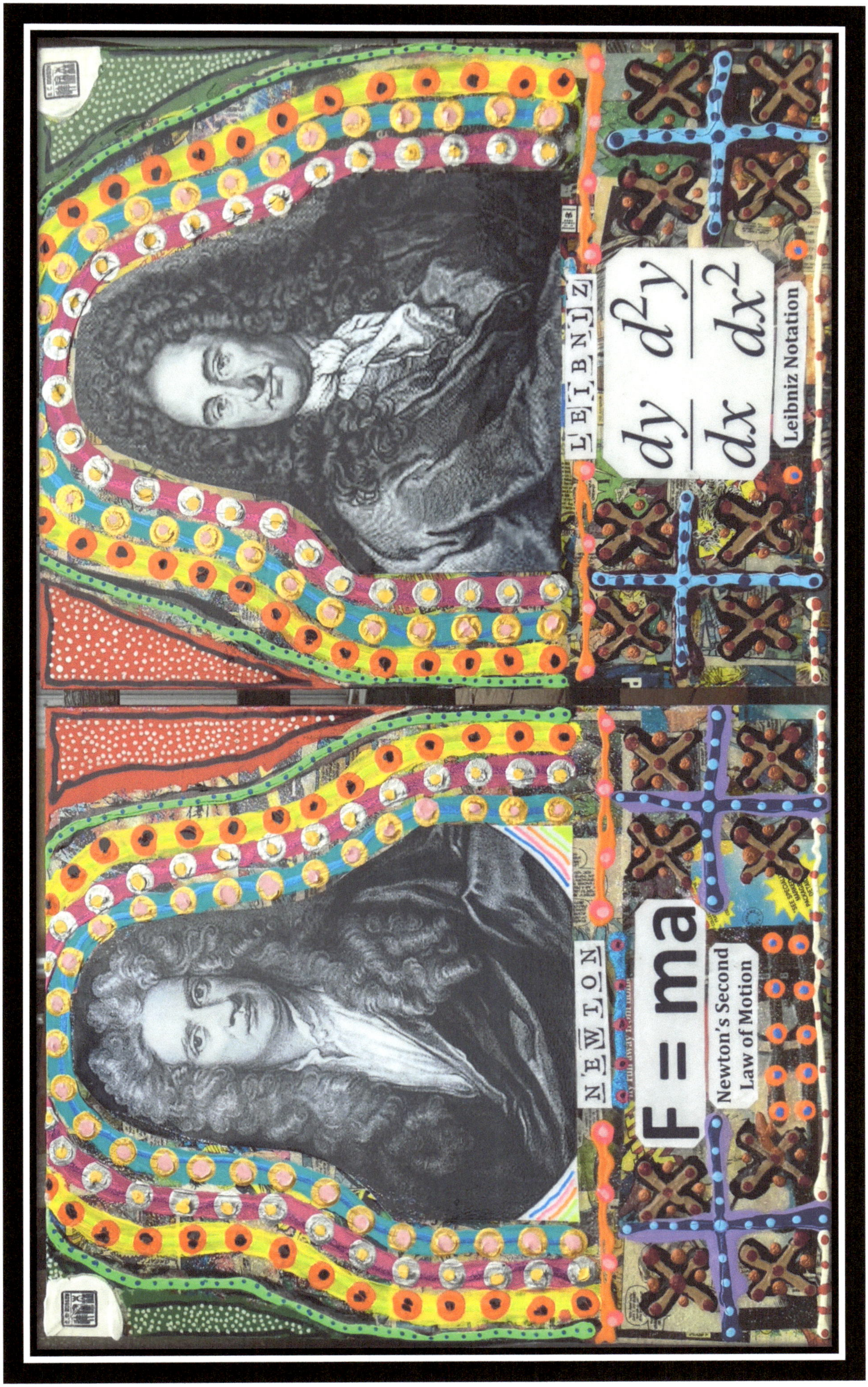
NEWTON
F = ma
Newton's Second Law of Motion
LEIBNIZ
$\frac{dy}{dx}$ $\frac{d^2y}{dx^2}$
Leibniz Notation

The Family Bernoulli (1623-1789)
Swiss

Jacob, Johann and Daniel are the most important Bernoullis. Jacob and Johann were brothers, and Daniel was one of Johann's sons, Jacob's nephew. Jacob and Johann were some of the first people to engage with the calculus of Newton and Leibniz. Jacob also studied matters of probability. Daniel is known for Bernoulli's Principle, a key concept in fluid dynamics that is important in aeronautics and in the carburetor of the internal combustion engine.

The Logarithmic Spiral

Jacob Bernoulli (1655-1705)

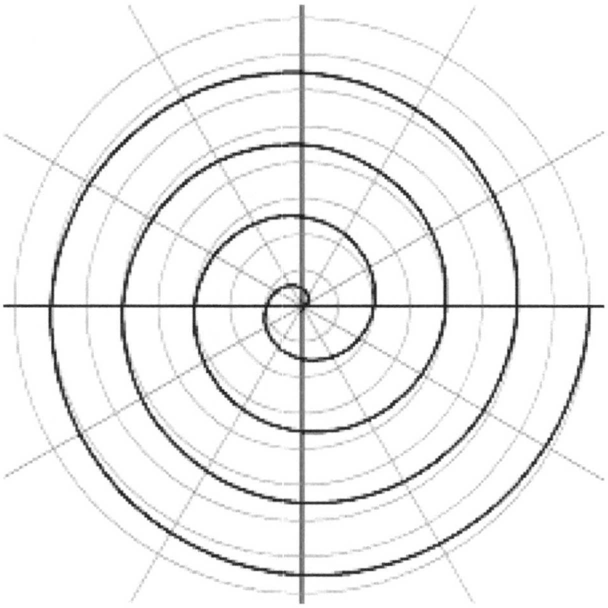

Bernoulli's Principle

Daniel Bernoulli (1700-1782)

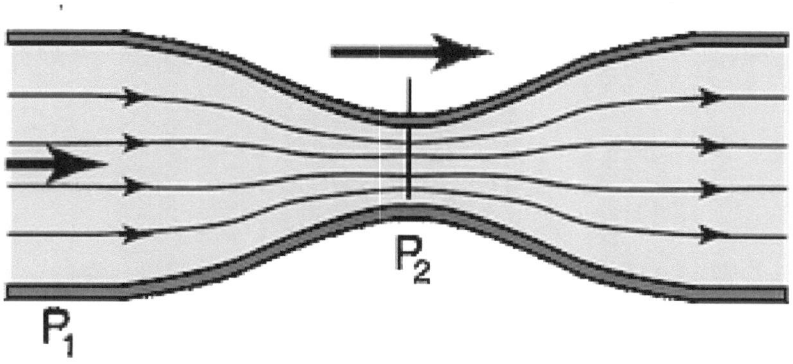

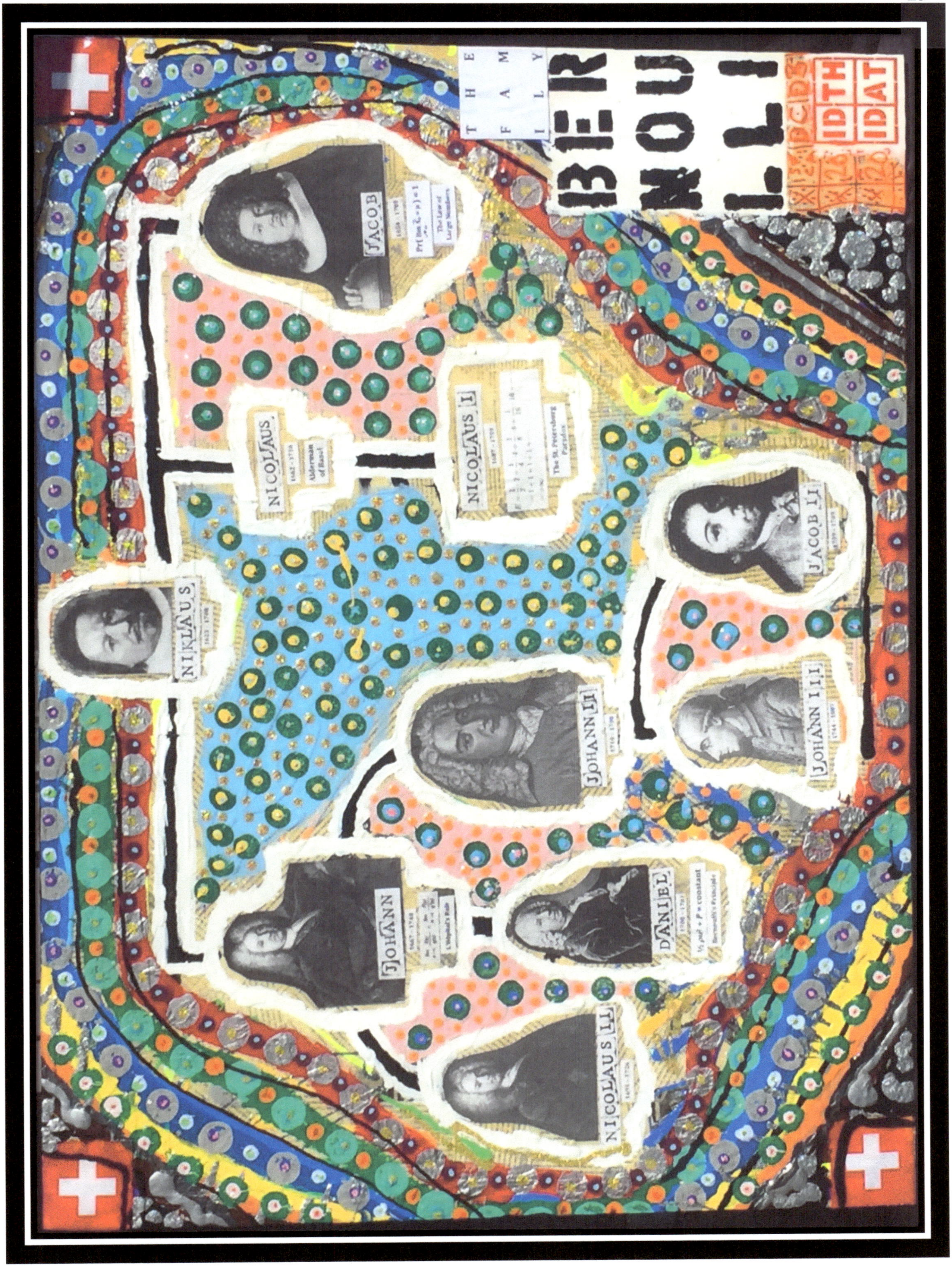
THE FAMILY
BERNOULLI
JACOB
The Law of Large Numbers
NICOLAUS
Alderman of Basel
NICOLAUS I
The St. Petersburg Paradox
NIKLAUS
JACOB II
JOHANN II
JOHANN III
JOHANN
DANIEL
NICOLAUS II

Pierre-Simon Laplace (1749-1827)
French

Laplace was a French aristocrat who taught a young Napoleon. He is a towering figure in European science. He was interested not only in mathematics and astronomy, but also ocean tides and the speed of sound, among other things. A great polymath.

The Laplace Transform

$$\mathscr{L}\{f(t)\} = F(s)$$

Laplace Distribution

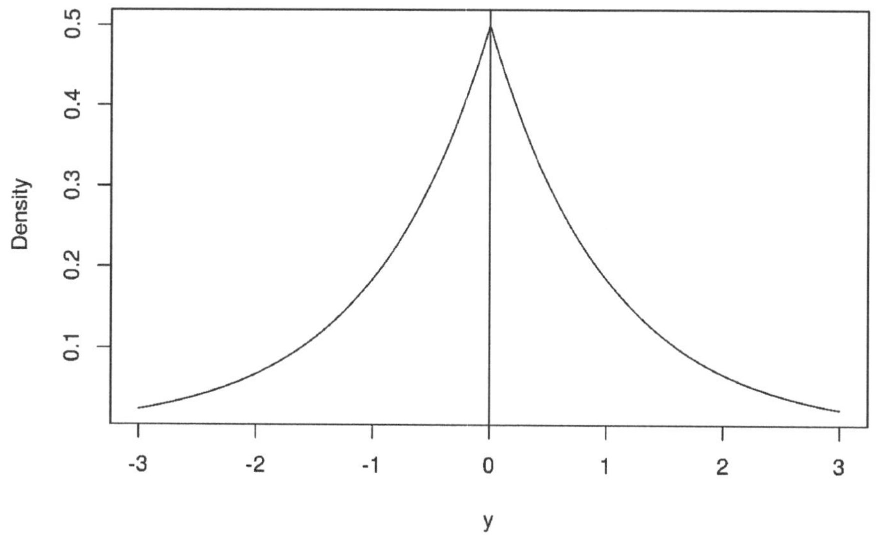

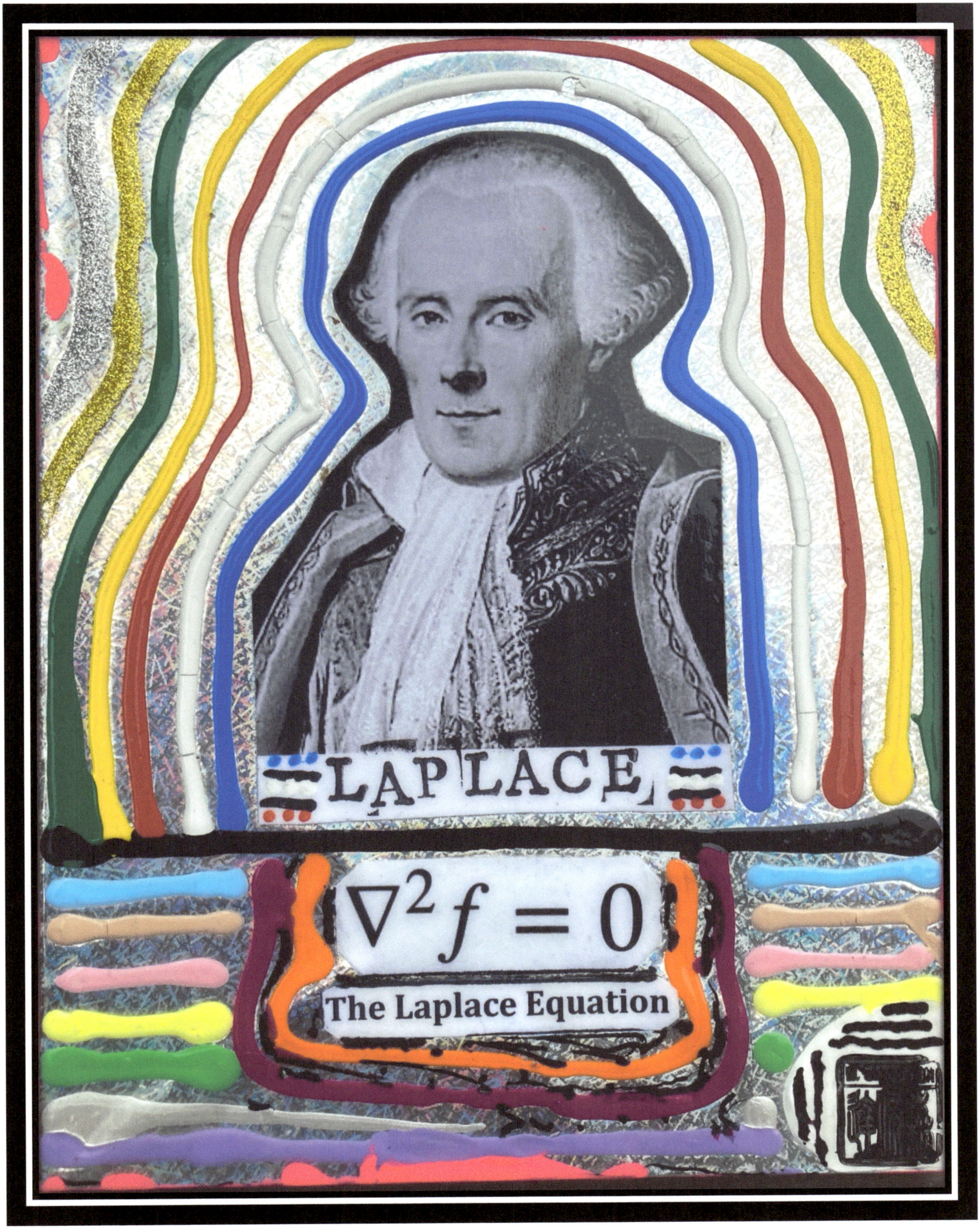
LAPLACE
$\nabla^2 f = 0$
The Laplace Equation

Carl Friedrich Gauss (1777-1855) and
Leonhard Euler (1707-1783)
German (Gauss) and Swiss (Euler)

Gauss and Euler both contributed to mathematics and to physics, but Gauss is known more for physics and Euler more for mathematics. These two men are considered by many to be the two greatest mathematicians in history.

Gauss's Heptadecagon

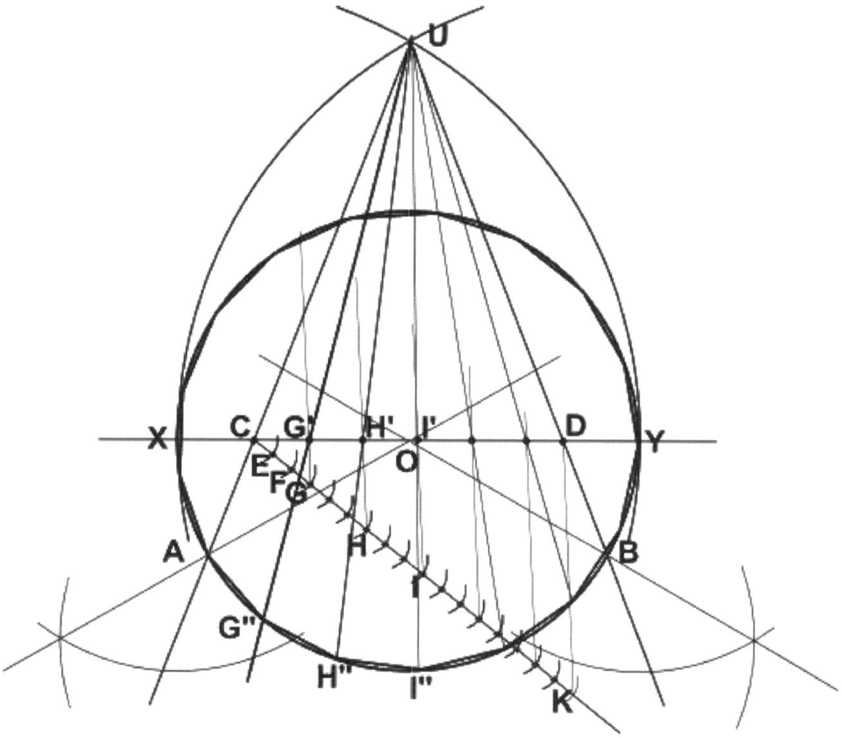

Euler's Number

The Base of the Natural Logarithm

$$e = \sum_{n=0}^{\infty} \frac{1}{n!} = \frac{1}{1} + \frac{1}{1} + \frac{1}{1 \cdot 2} + \frac{1}{1 \cdot 2 \cdot 3} + \cdots$$

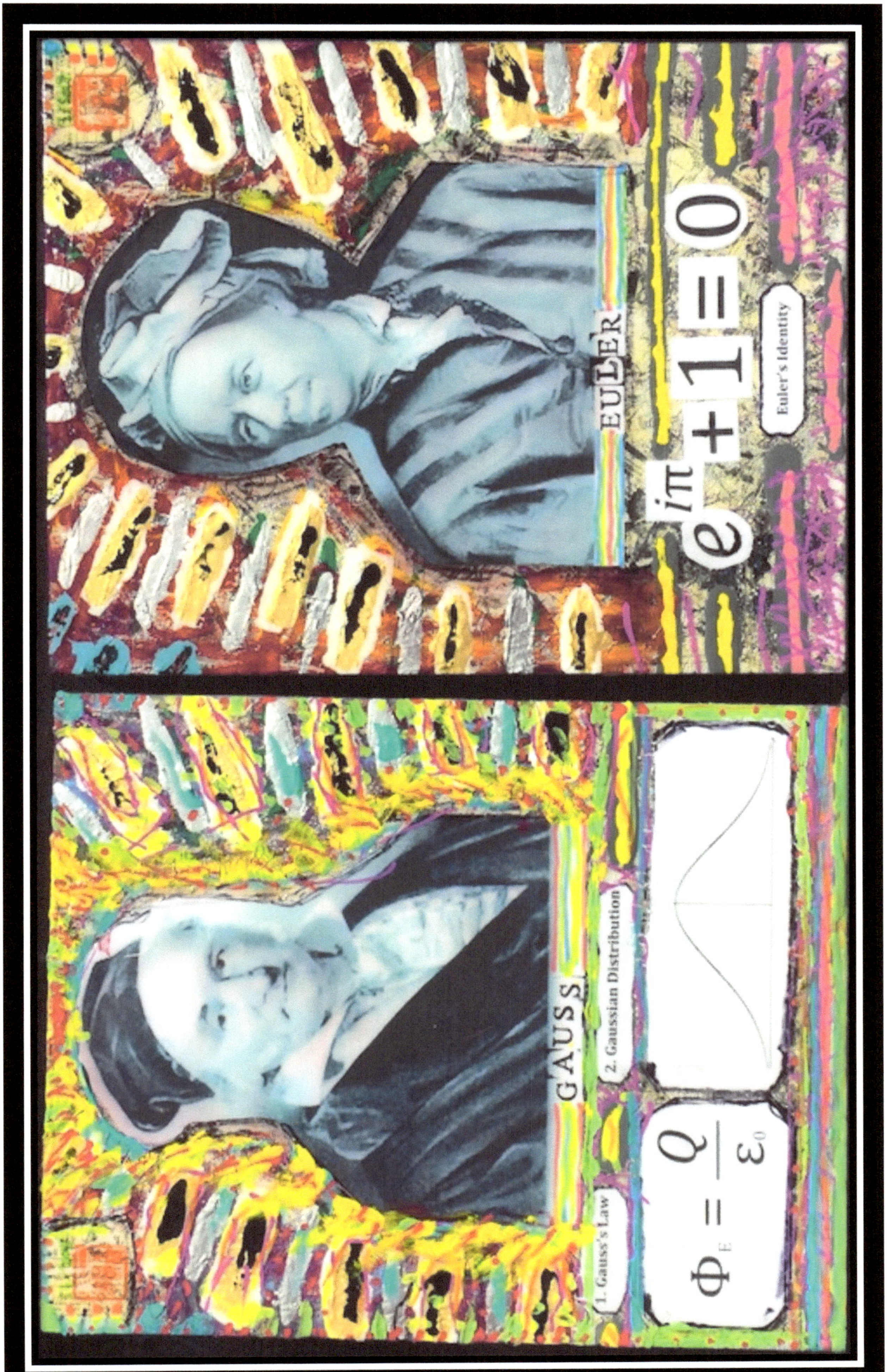
GAUSS
1. Gauss's Law
2. Gaussian Distribution
$\Phi_E = \frac{Q}{\varepsilon_0}$
EULER
$e^{i\pi} + 1 = 0$
Euler's Identity

Augustin-Louis Cauchy (1789-1857)
French

Cauchy basically single-handedly invented the mathematical field of complex analysis, which is the study of functions with complex variables, that is, variables of the form a + b*i*, *i* being the imaginary number equal to the square root of -1. It so happens that this subject is vital both to mathematics and to physics. Cauchy also made many contributions to engineering. His is one of the 42 names memorialized on the Eiffel Tower.

Cauchy-Riemann Equations

$$\frac{\partial u}{\partial x} = \frac{\partial v}{\partial y} \qquad \frac{\partial u}{\partial y} = -\frac{\partial v}{\partial x}$$

Cauchy-Euler Equation

$$ax^2y'' + bxy' + cy = 0$$

Cauchy Distribution

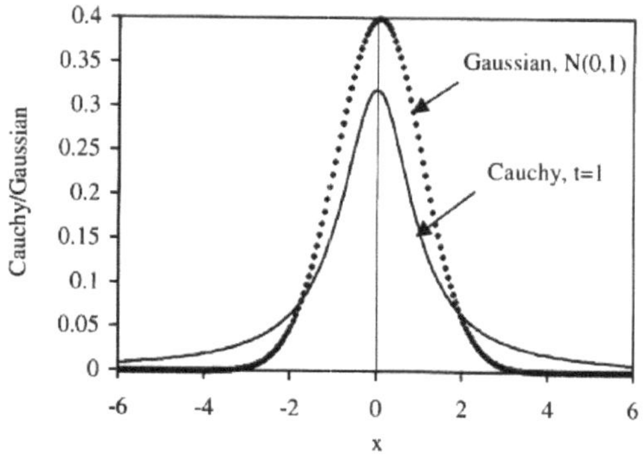

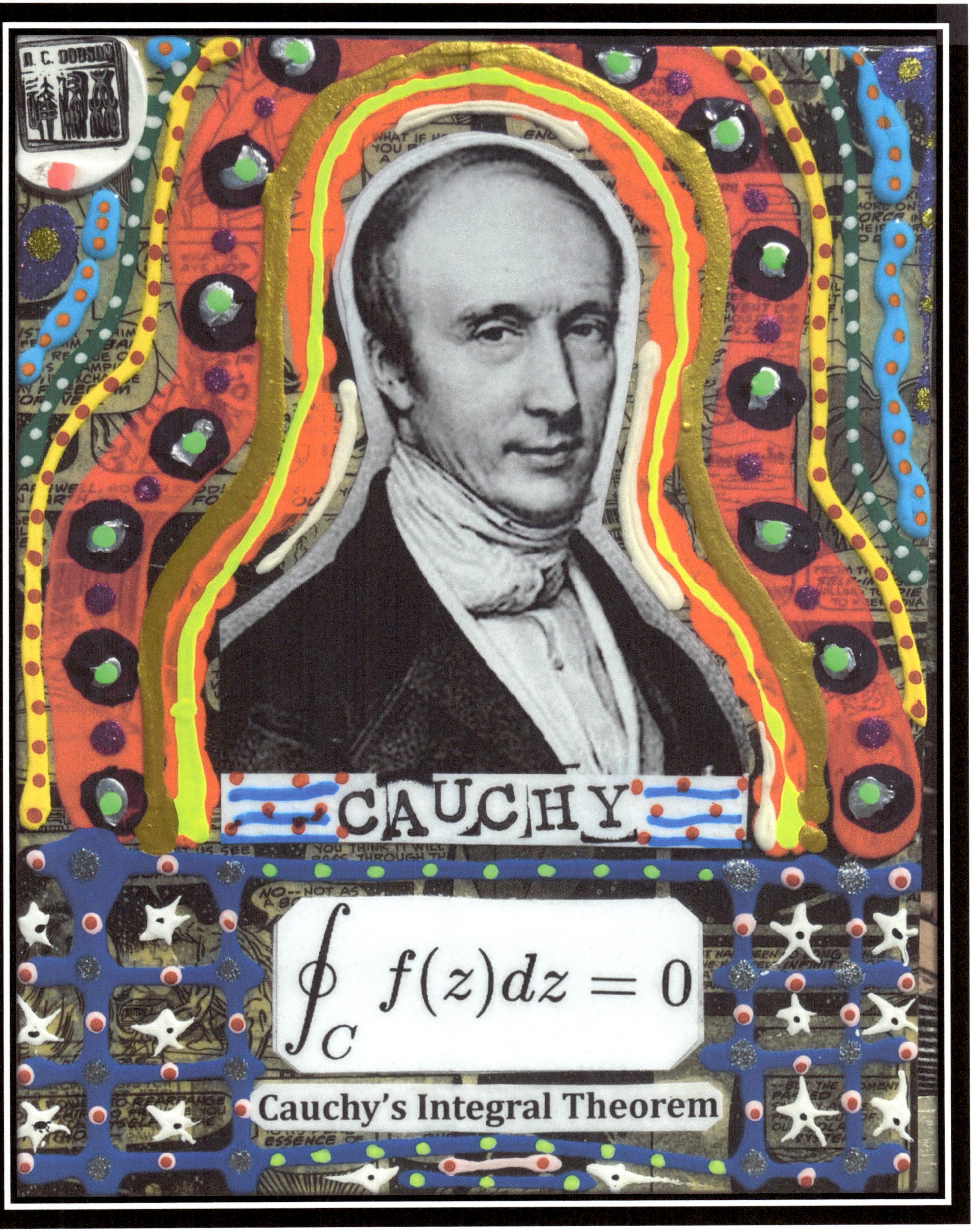
R. C. DOBSON
CAUCHY
$$\oint_C f(z)dz = 0$$
Cauchy's Integral Theorem

Niels Henrik Abel (1803-1829)
Norwegian

Abel was a Norwegian mathematician in the 19th century. He contributed to multiple areas of mathematics before dying of tuberculosis at age 26. The highly prestigious Abel Prize in mathematics, awarded annually by the King of Norway, is named after him.

Abel Equation

$$\xi = \int f_3(x) E^2 \, dx,$$

$$u = \left(y + \frac{f_2(x)}{3 f_3(x)} \right) E^{-1},$$

$$E = \exp\left(\int \left(f_1(x) - \frac{f_2^2(x)}{3 f_3(x)} \right) dx \right)$$

Abel's Theorem

If y_1 and y_2 are 2 solutions to the equation

$$y'' + p(x)y' + q(x)y = 0$$

Then $W(y_1, y_2) = ce^{-\int p(t)dt}$

Ex:

$$t^2 y'' + 2t^3 y' - t^{-2} y = 0$$

$$y'' + 2ty' - \frac{1}{t^4} y = 0$$

$$W(y_1, y_2) = ce^{-\int 2t\,dt} = ce^{-t^2}$$

$$W(y_1, y_2) = \begin{vmatrix} y_1 & y_2 \\ y_1' & y_2' \end{vmatrix} = y_1 y_2' - y_2 y_1'$$

$$= ce^{-t^2}$$

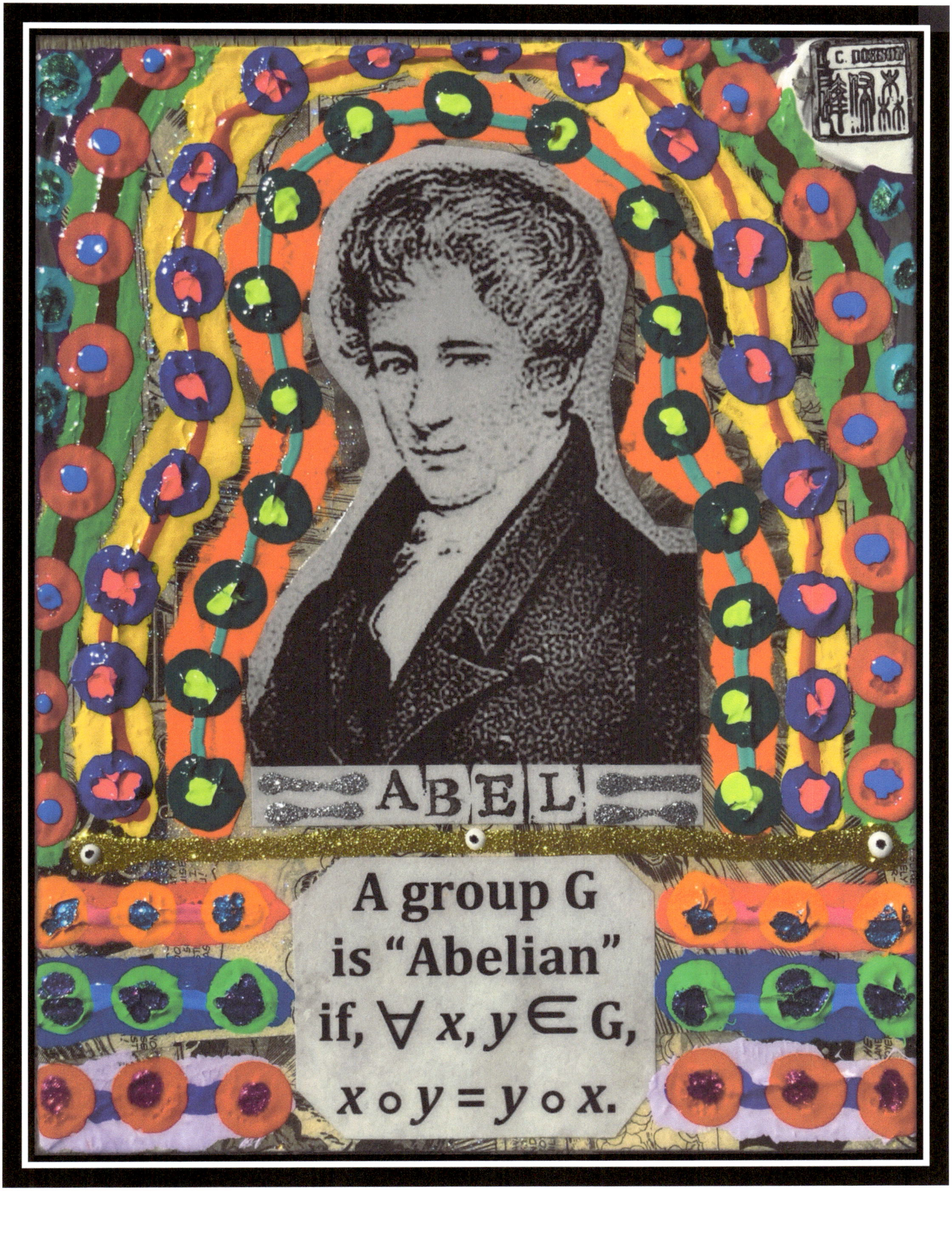
ABEL
A group G
is “Abelian”
if, ∀ x, y ∈ G,
x ∘ y = y ∘ x.

William Rowan Hamilton (1805-1865)
Irish

Hamilton is primarily known for inventing/discovering quaternions, a system of four-dimensional numbers. He was also a noted astronomer and made numerous other contributions. He spent his professional life at Trinity College in Dublin.

The Differential Operator

$$\nabla = \frac{\partial}{\partial x} i + \frac{\partial}{\partial y} j + \frac{\partial}{\partial z} k$$

Hamilton himself used a slightly different symbol. The modern symbol, seen here, is known as "del" or "nabla", the latter coming from a Greek word for "harp".

Hamilton's Equations

$$\frac{\mathrm{d}\boldsymbol{q}}{\mathrm{d}t} = \frac{\partial \mathcal{H}}{\partial \boldsymbol{p}}$$

$$\frac{\mathrm{d}\boldsymbol{p}}{\mathrm{d}t} = -\frac{\partial \mathcal{H}}{\partial \boldsymbol{q}}$$

Here, $\boldsymbol{q}$ (coordinate or position) and $\boldsymbol{p}$ (momentum) are related using the Hamiltonian $\mathcal{H}$. $\mathcal{H}$ is basically the total energy of a system expressed as a function of $\boldsymbol{p}$ and $\boldsymbol{q}$, which are themselves functions of time. What specific function $\mathcal{H}$ actually is varies depending on the application at hand.

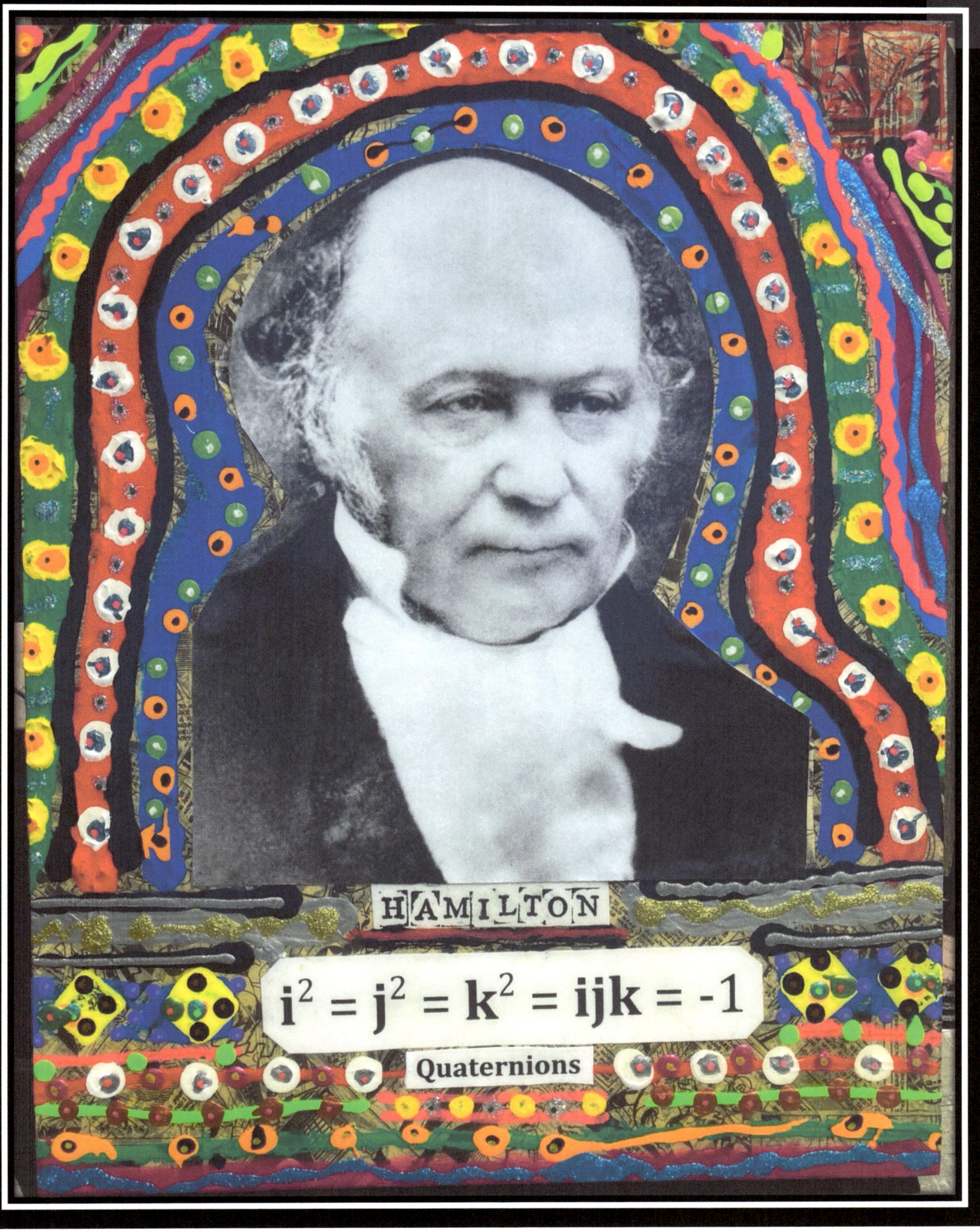
HAMILTON
$i^2 = j^2 = k^2 = ijk = -1$
Quaternions

George Stokes (1819-1903)
Irish

George Stokes was an Irish physicist and mathematician. He spent most of his life and career at Cambridge University in England. He is perhaps best known today for the Navier-Stokes Equations, a theory about fluid dynamics. Finding certain solutions to these equations is one of the unsolved Millennium Prize problems, for which there is an outstanding prize of one million dollars, first offered in 2000 by the Clay Mathematics Institute.

Navier-Stokes Equations

$$\rho g_x - \frac{\partial p}{\partial x} + \mu\left(\frac{\partial^2 u}{\partial x^2} + \frac{\partial^2 u}{\partial y^2} + \frac{\partial^2 u}{\partial z^2}\right) = \rho \frac{Du}{Dt}$$

$$\rho g_y - \frac{\partial p}{\partial y} + \mu\left(\frac{\partial^2 v}{\partial x^2} + \frac{\partial^2 v}{\partial y^2} + \frac{\partial^2 v}{\partial z^2}\right) = \rho \frac{Dv}{Dt}$$

$$\rho g_z - \frac{\partial p}{\partial z} + \mu\left(\frac{\partial^2 w}{\partial x^2} + \frac{\partial^2 w}{\partial y^2} + \frac{\partial^2 w}{\partial z^2}\right) = \rho \frac{Dw}{Dt}$$

$$\frac{\partial \rho}{\partial t} + \nabla \cdot (\rho \overline{V}) = 0$$

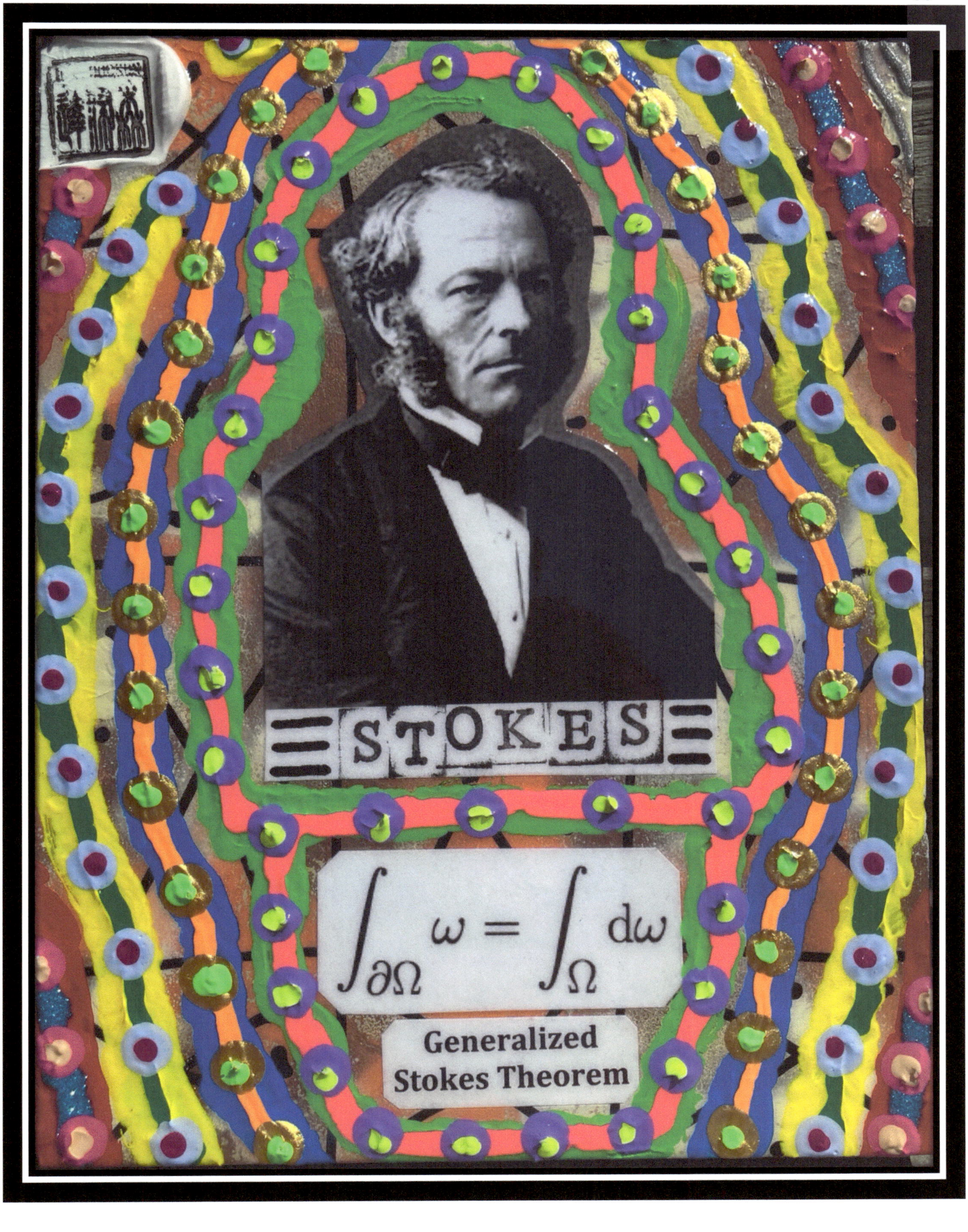
STOKES
$\int_{\partial\Omega} \omega = \int_{\Omega} d\omega$
Generalized
Stokes Theorem

James Clerk Maxwell (1831-1879) and
Michael Faraday (1791-1867)
Scottish (Maxwell) and English (Faraday)

These two Britons managed to redefine electricity, magnetism, and light as several manifestations of the same phenomenon: electromagnetism. Faraday in particular demonstrated how this phenomenon could be put to use for mankind.

The Electromagnetic Wave

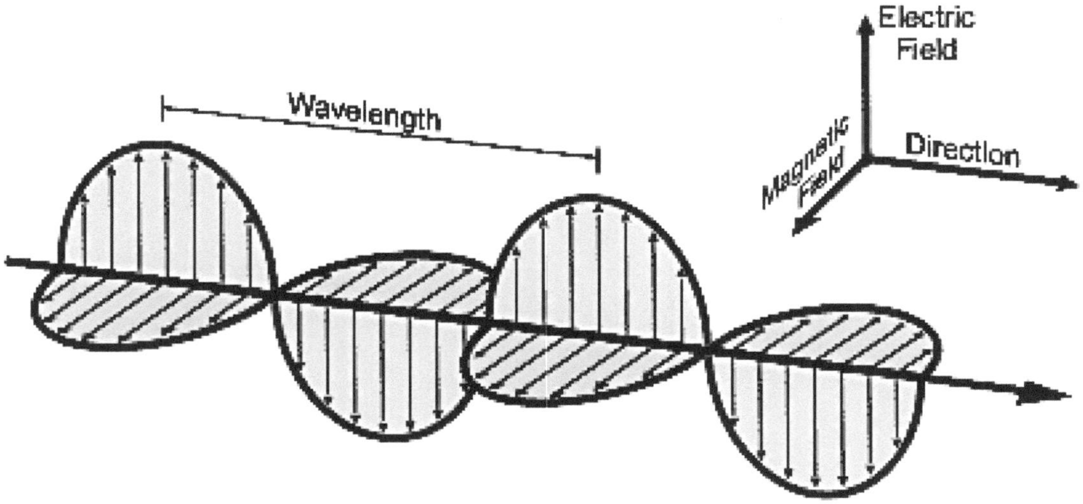

The Electromagnetic Spectrum

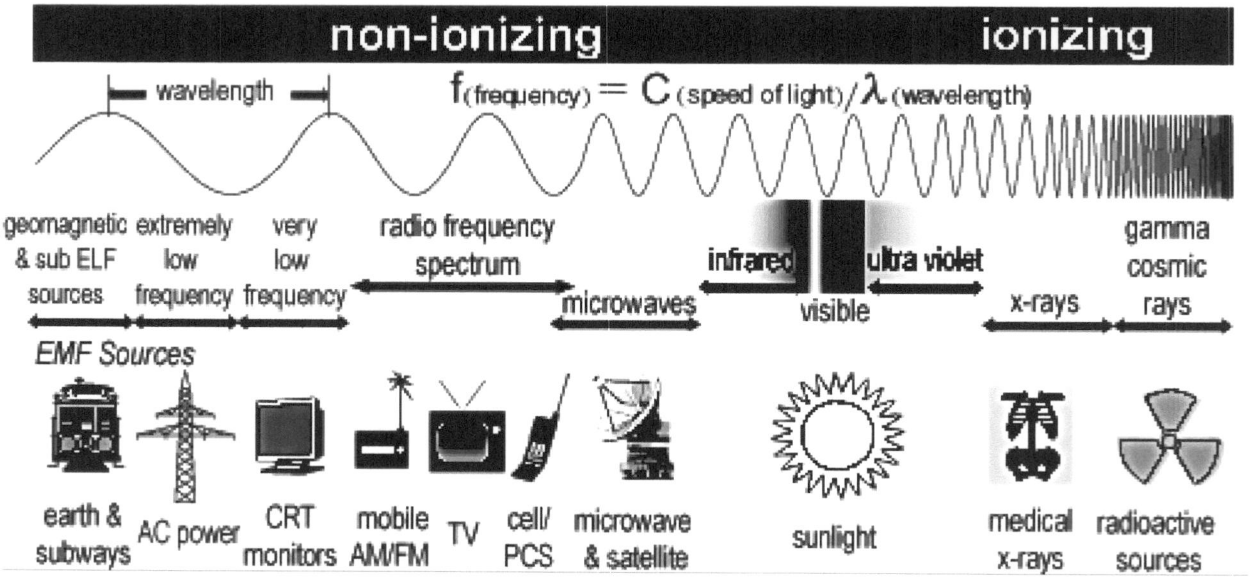

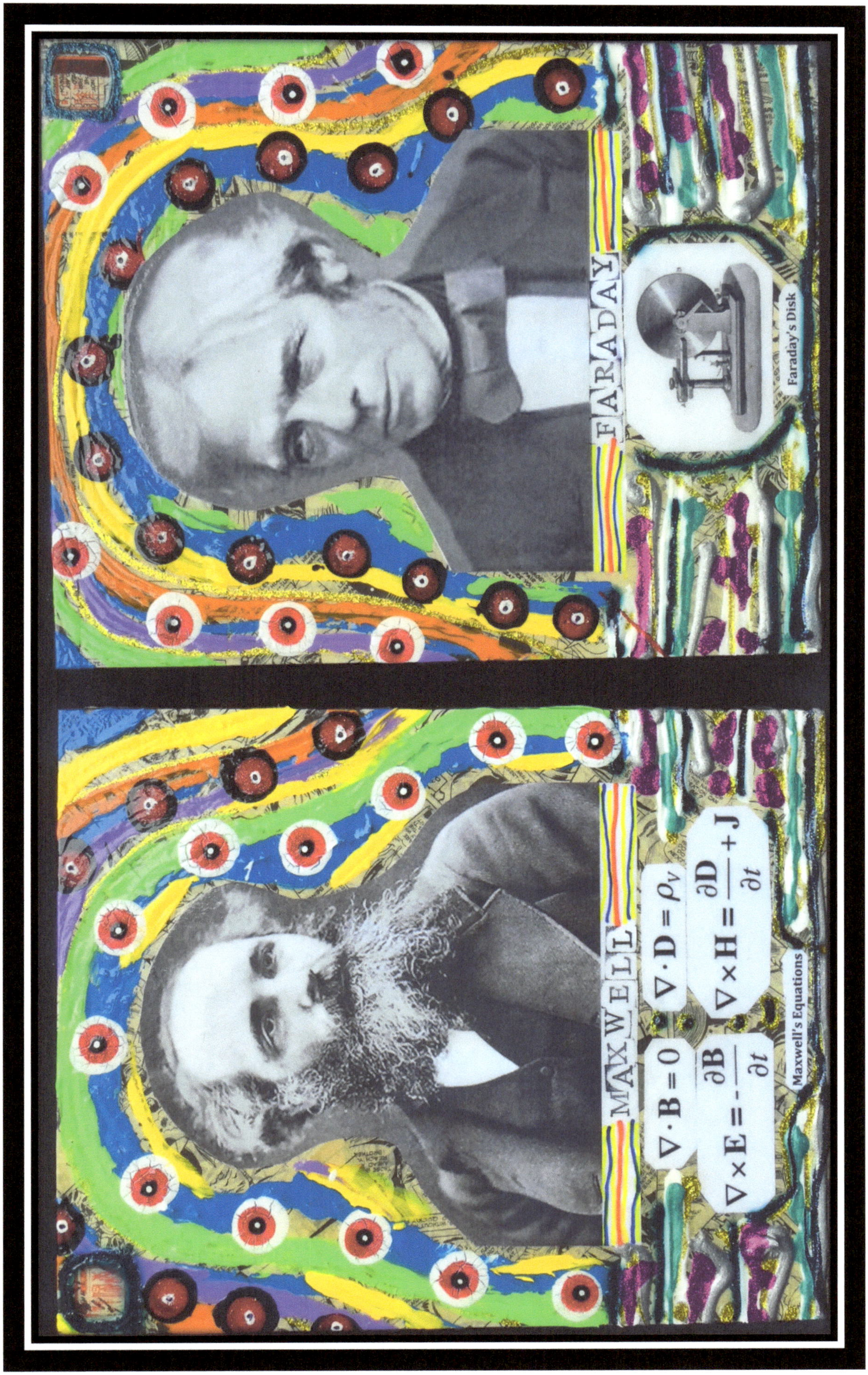
FARADAY
Faraday's Disk
MAXWELL
$\nabla \cdot \mathbf{B} = 0$
$\nabla \cdot \mathbf{D} = \rho_V$
$\nabla \times \mathbf{E} = -\frac{\partial \mathbf{B}}{\partial t}$
$\nabla \times \mathbf{H} = \frac{\partial \mathbf{D}}{\partial t} + \mathbf{J}$
Maxwell's Equations

Bernhard Riemann (1826-1866)
German

The Riemann Hypothesis is one of the most pressing topics in mathematics today. It is one of the Millennium Prize Problems, for which there is an outstanding million-dollar prize. To prove it would establish a major connection between number theory and complex analysis by showing that complex-plane geometry can tell us much about the sequence of the prime numbers.

A Riemannian Manifold

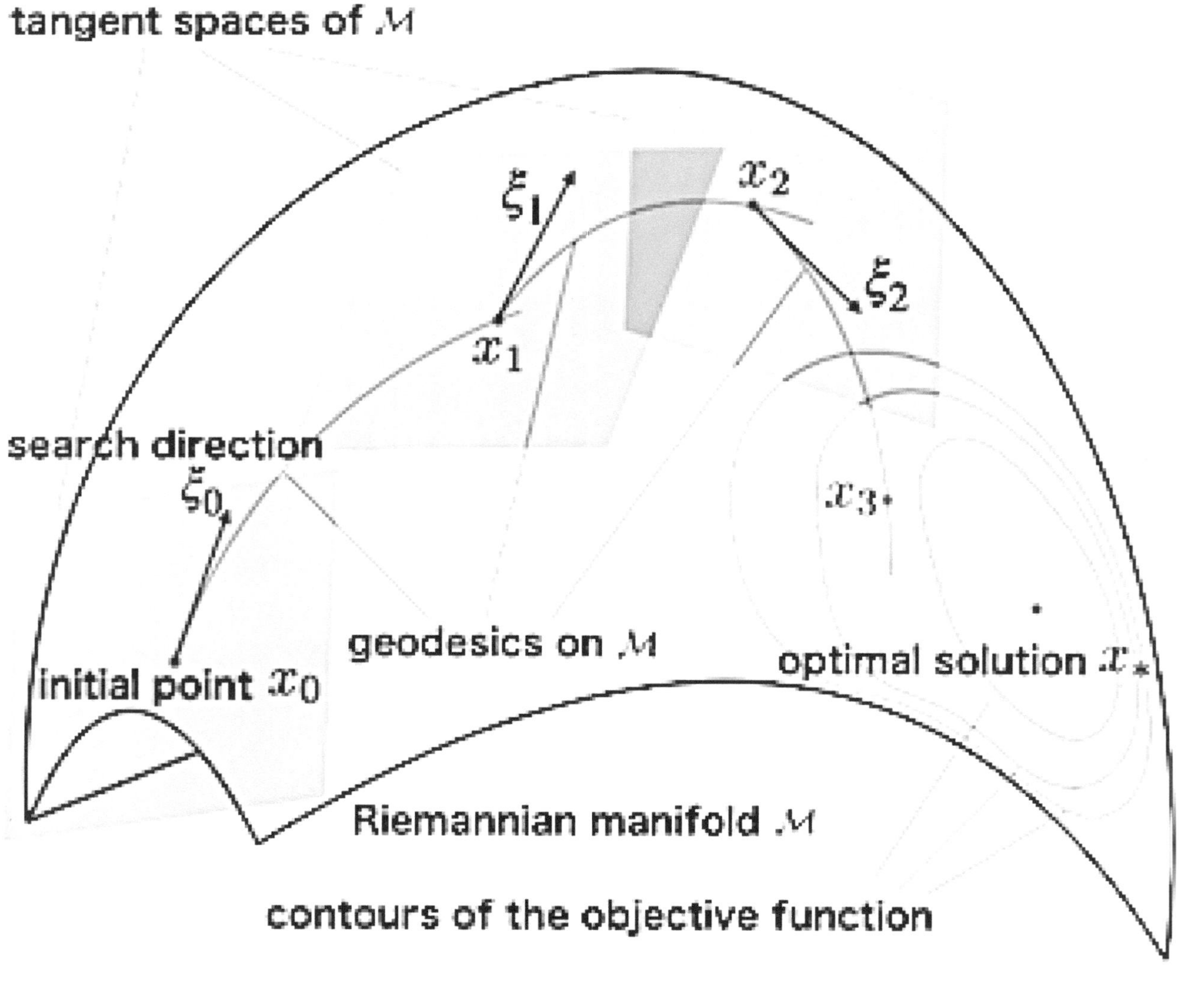

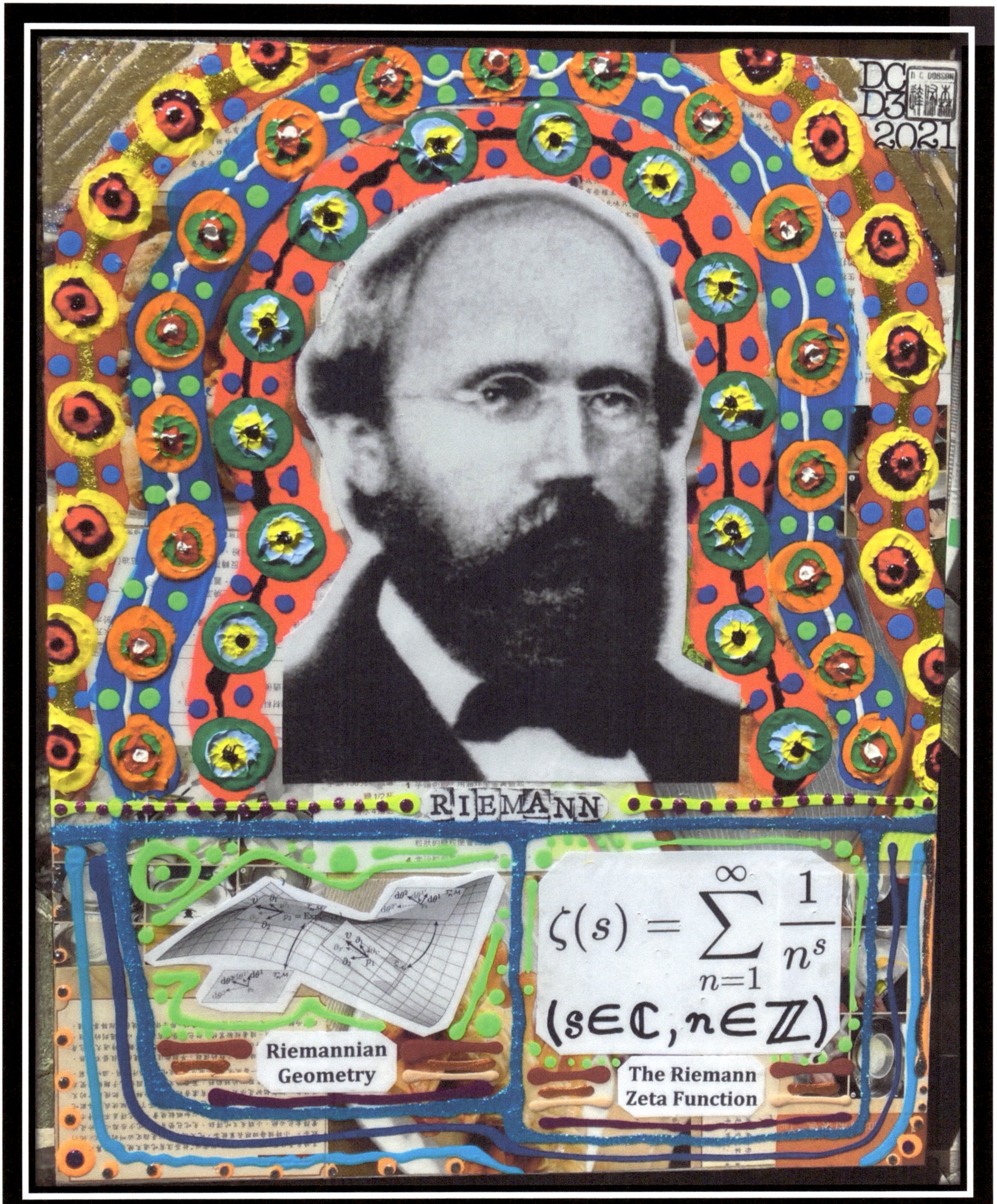
DC D3 2021
RIEMANN
$\zeta(s) = \sum_{n=1}^{\infty} \frac{1}{n^s}$
$(s \in \mathbb{C}, n \in \mathbb{Z})$
Riemannian Geometry
The Riemann Zeta Function

William Kingdon Clifford (1845-1879)
English

Clifford Algebra is the basis for the current mathematical field of Geometric Algebra. To establish a real unity between geometry and algebra is a pressing topic in mathematics today.

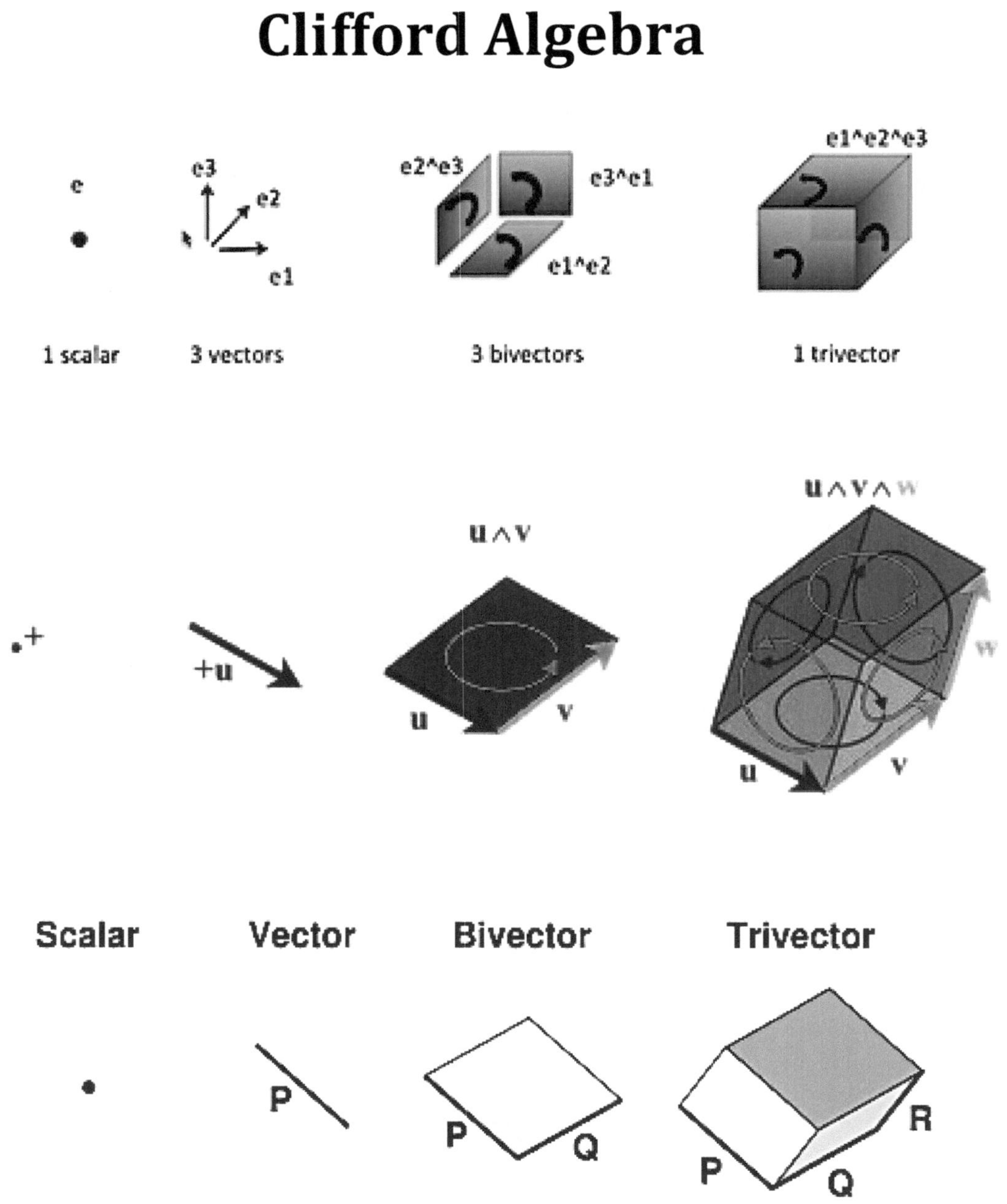

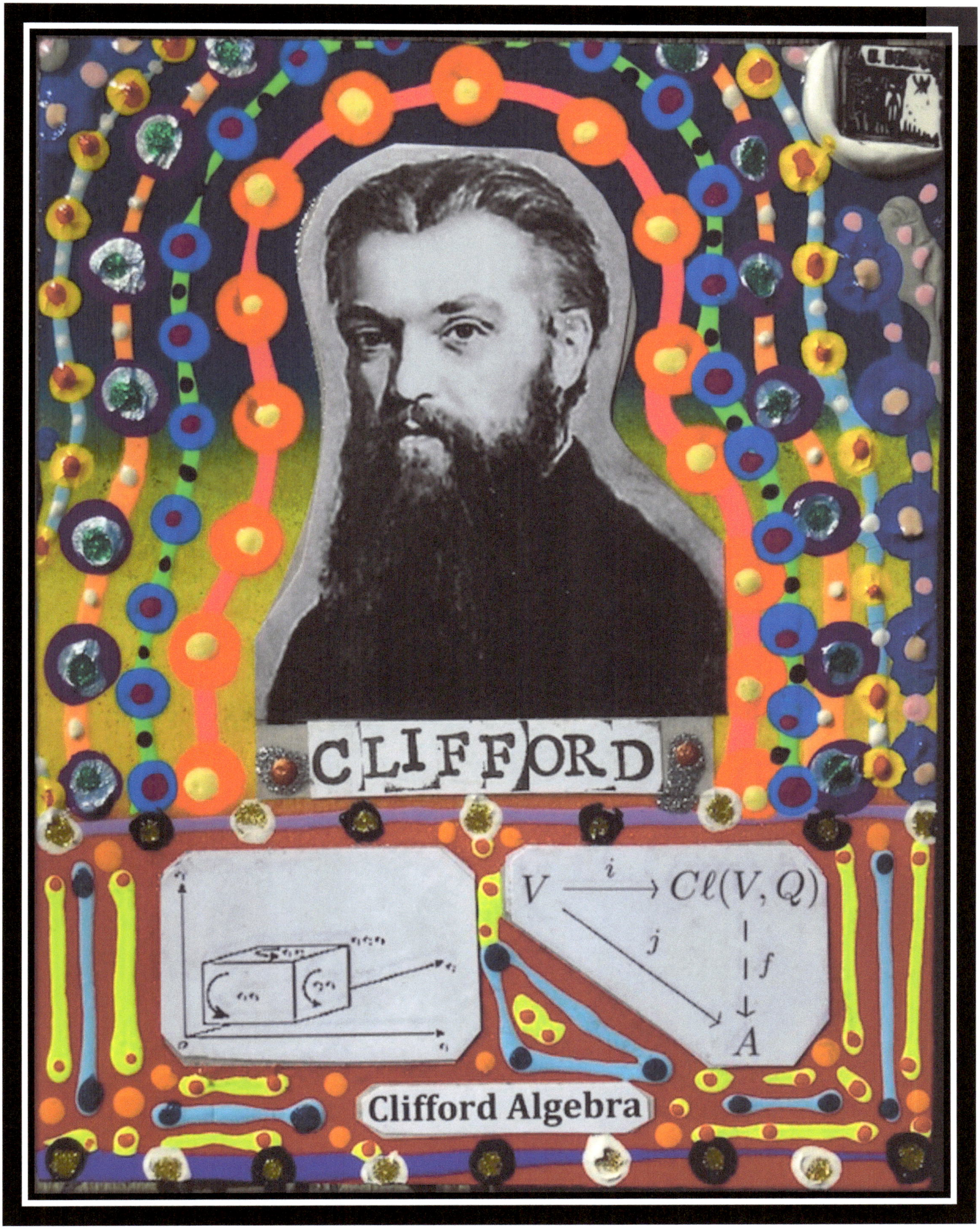
CLIFFORD
$V \xrightarrow{i} C\ell(V,Q)$
j
f
A
Clifford Algebra

Oliver Heaviside (1850-1925)
English

Heaviside did not so much make original discoveries as he refined the scientific work of the 19th century and found new ways to apply previous findings. The present form of Maxwell's Equations is due to Heaviside. The Telegrapher's Equations, seen here, describe current and voltage in telegraph wires.

Heaviside's Formulation of Maxwell's Equations

$\nabla \cdot \mathbf{D} = \rho$	(1)	Gauss' Law
$\nabla \cdot \mathbf{B} = 0$	(2)	Gauss' Law for magnetism
$\nabla \times \mathbf{E} = -\frac{\partial \mathbf{B}}{\partial t}$	(3)	Faraday's Law
$\nabla \times \mathbf{H} = \frac{\partial \mathbf{D}}{\partial t} + \mathbf{J}$	(4)	Ampère-Maxwell Law

The Ionosphere

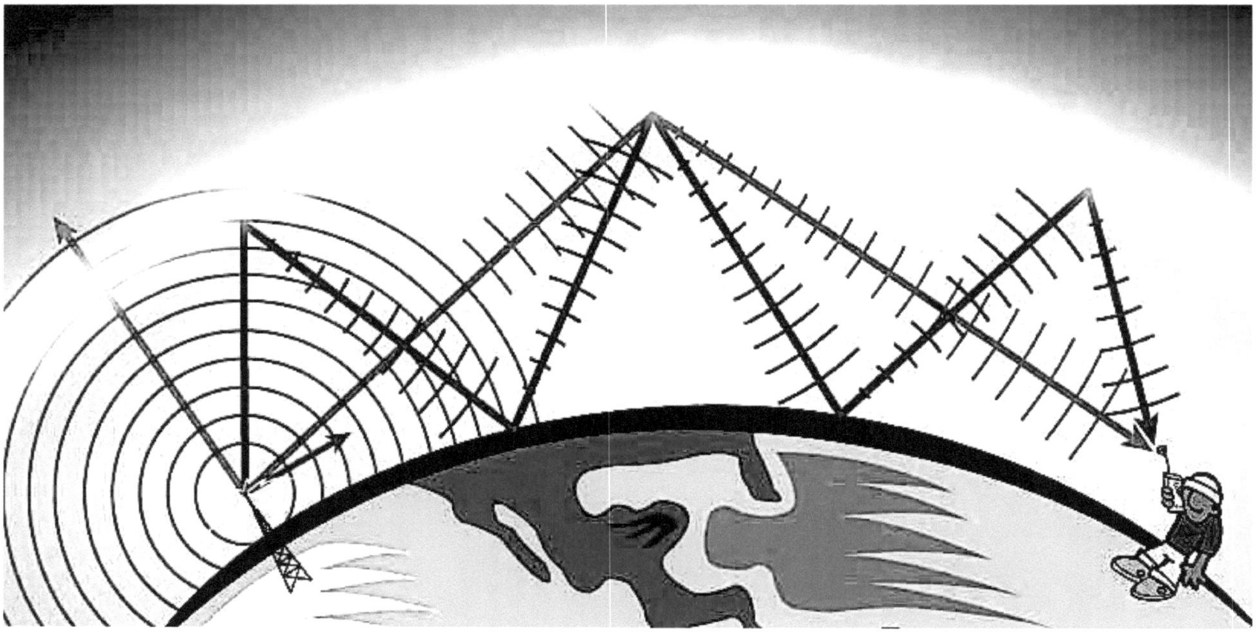

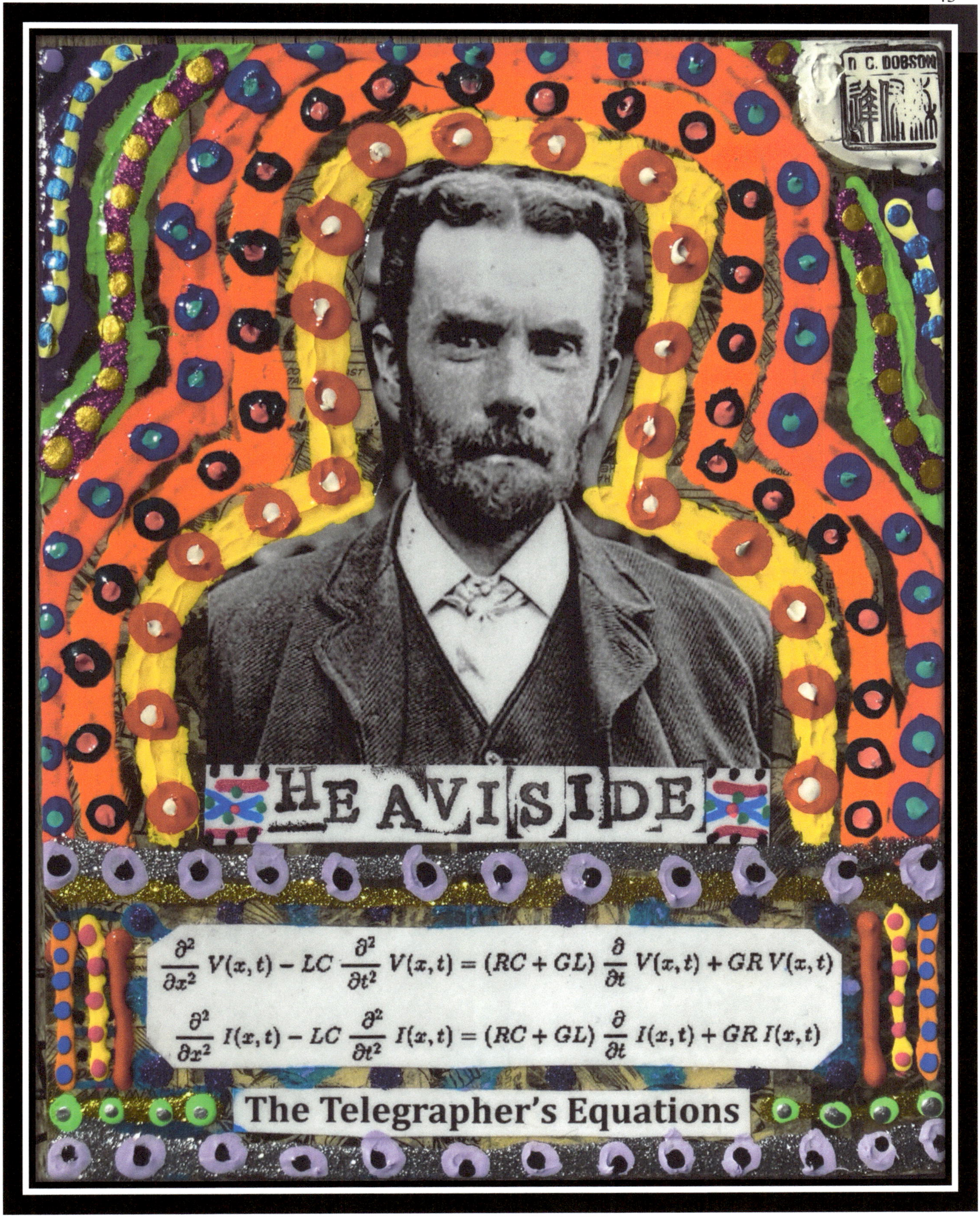

D. C. DOBSON
HEAVISIDE
$$\frac{\partial^2}{\partial x^2} V(x,t) - LC \frac{\partial^2}{\partial t^2} V(x,t) = (RC + GL) \frac{\partial}{\partial t} V(x,t) + GR\, V(x,t)$$
$$\frac{\partial^2}{\partial x^2} I(x,t) - LC \frac{\partial^2}{\partial t^2} I(x,t) = (RC + GL) \frac{\partial}{\partial t} I(x,t) + GR\, I(x,t)$$
The Telegrapher's Equations

Henri Poincaré (1854-1912)
French

Poincaré is something of a French folk hero. The Lycée Henri Poincaré can be found in Nancy in northern France. Poincaré was an intellectually voracious polymath. He has been called the Last Universalist, since he was one of the last to see mathematics, physics, and engineering as one unified discipline. Poincaré's Sphere is a way to depict the different ways light can be polarized. It's not his most important work, but it comes with nice graphics.

Poincaré Group

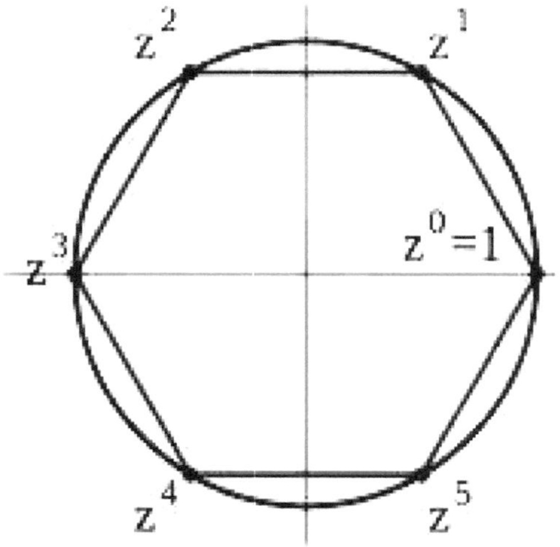

Poincaré Duality

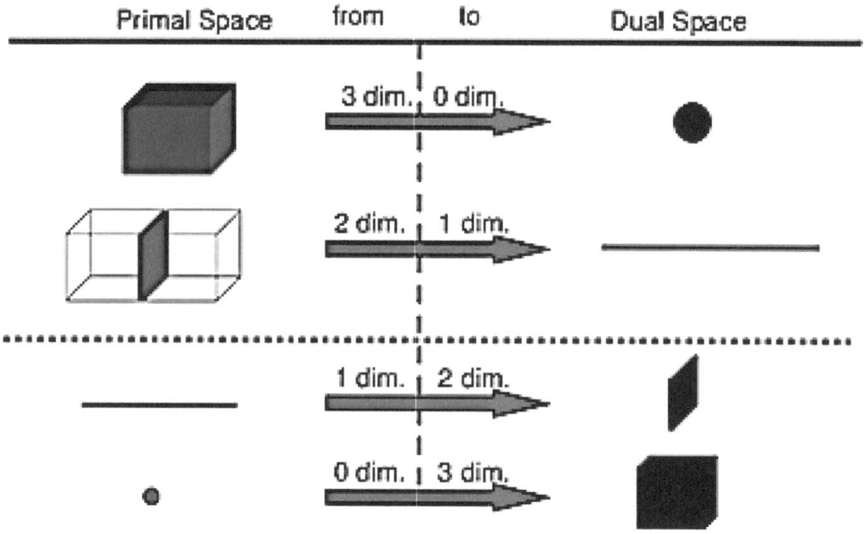

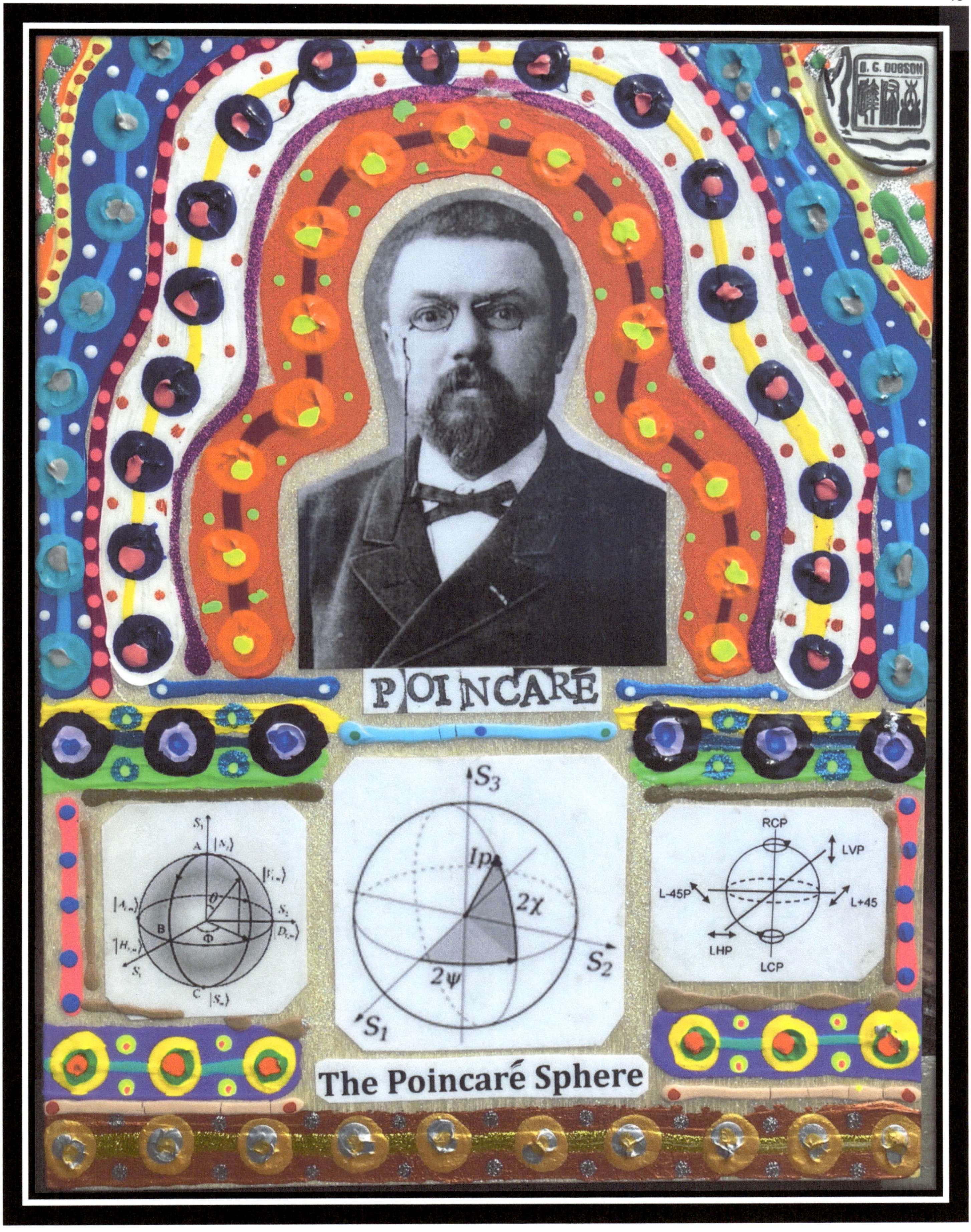
B. G. DOBSON
POINCARÉ
S3
Ip
2χ
S2
2ψ
S1
RCP
LVP
L-45P
L+45
LHP
LCP
The Poincaré Sphere

Max Planck (1858-1947)
German

Planck, a lifelong professional theoretical physicist, is known as the founder of quantum theory. His constant lies at the heart of particle physics. The word "quantum" means "physical unit." The quanta of quantum theory are the very particles themselves. By fixing a certain value that conforms with experiment, Planck drew a line in the proverbial sand. When combined with Bohr's model of the atom, which involves electrons making discrete jumps between energy levels, Planck's Constant makes a lot of sense.

Planck's Constant

h

$= 6.62 \times 10^{-34}$

m^2kg/s

Planck's Law

$$E(\lambda, T) = \frac{2hc^2}{\lambda^5} * \frac{1}{e^{\left(\frac{hc}{\lambda kT}\right)} - 1}$$

h = Planck's constant = $6.626 * 10^{-34}\ J * s$
c = speed of light = $2.997925 * 10^{8}\ m / sec$
λ = wavelength (m)
k = Boltzmann's constant = $1.381 * 10^{-23}\ J/K$
T = temperature (K)

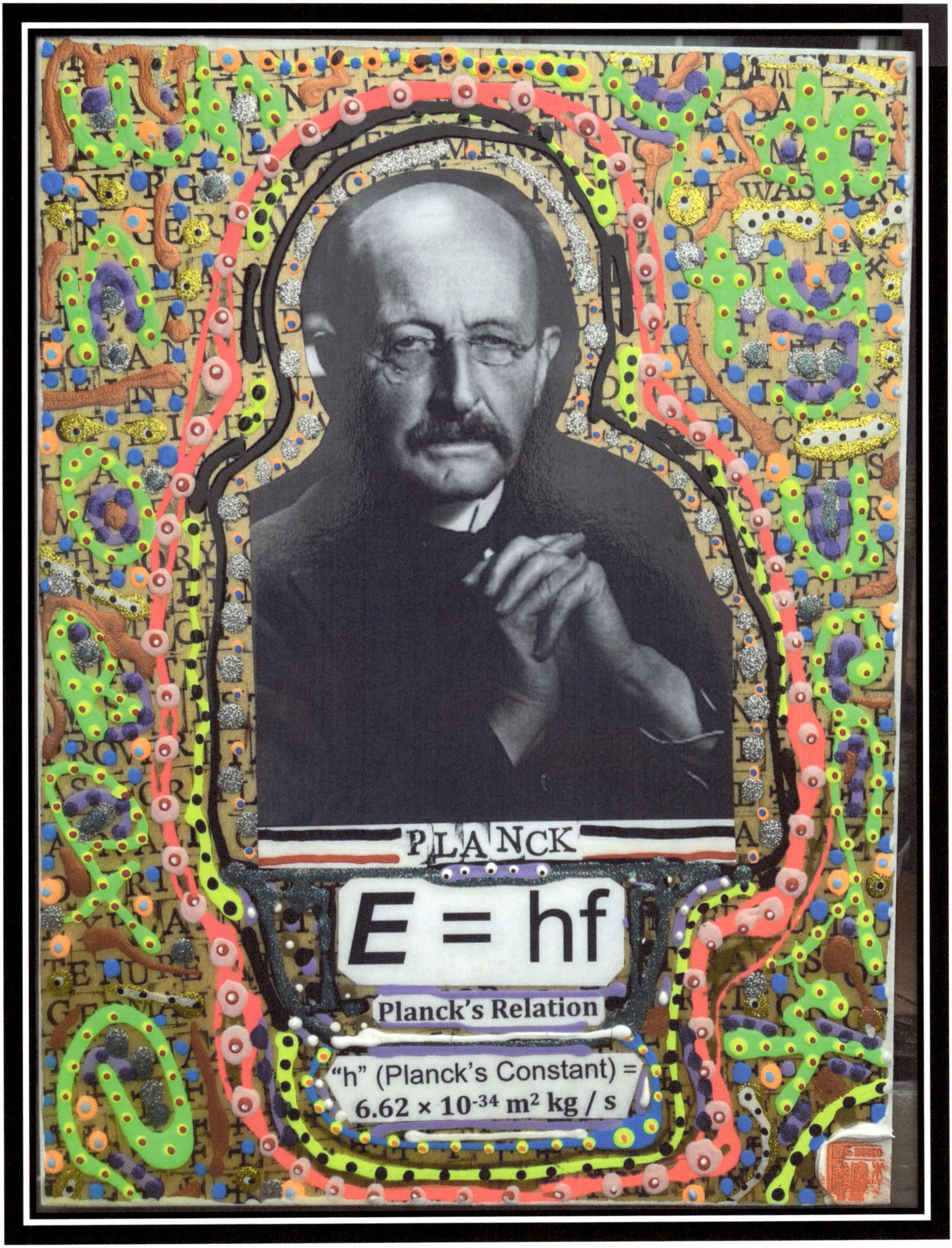
PLANCK
E = hf
Planck's Relation
"h" (Planck's Constant) =
6.62 × 10-34 m2 kg / s

David Hilbert (1862-1943) and Georg Cantor (1845-1918)
German

"We must know! We will know!" Hilbert set the course of 20th-century mathematics with a list of 19 problems that he presented at a conference in Paris in 1900. Hilbert's most famous contribution is probably Hilbert Space, a generalization of flat Euclidean Space to higher dimensions. In the late 19th century, Cantor introduced the mathematical field of set theory, which tells us much about what numbers really are and how they interact. He also extensively studied the concept of infinity and infinite sets, showing, among other things, that infinity is not monolithic but comes in different varieties.

Some of the notation of modern set theory includes:

$\mathbf{N}$ = the set of natural numbers
$\mathbf{Q}$ = the set of rational numbers
$\mathbf{R}$ = the set of real numbers
$\mathbf{P}$ = the set of prime numbers
$\mathbf{Z}$ = the set of integers
$\mathbf{E}$ = the set of even integers
$\mathbf{O}$ = the set of odd integers

The symbol for showing that one set is a member of another set is $\subset$, so that:

$$\mathbf{P} \subset \mathbf{N} \subset \mathbf{Z} \subset \mathbf{Q}$$

The symbol for showing that one set is equivalent to another set is $\sim$, so that:

$$\mathbf{Q} \sim \mathbf{Z} \sim \mathbf{N} \sim \mathbf{P} \sim \mathbf{E} \sim \mathbf{O}$$

Curly brackets {} are used to show a collection of elements in a set:

$$\mathbf{A} = \{3,7,9,14\}$$

The number of members in a set is shown as | | or #:

$$\text{If } \mathbf{A} = \{3,7,9,14\}, \text{ then } |\mathbf{A}| = 4 \text{ or } \#\mathbf{A} = 4$$

The cardinality of infinite sets is denoted by $\aleph$:

$$|\mathbf{N}| = \aleph_0 \qquad |\mathbf{R}| = 2^{\aleph_0}$$

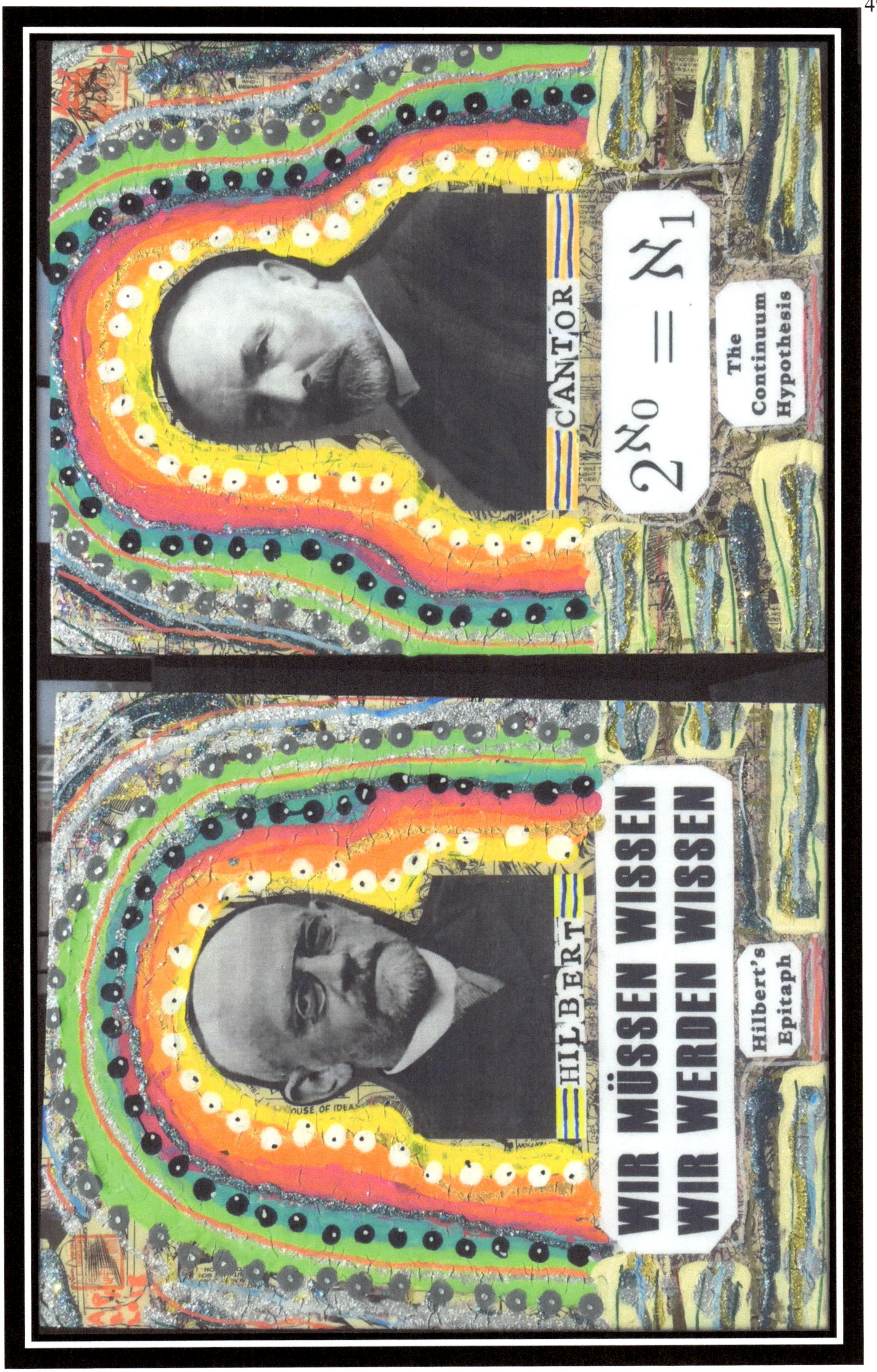
HILBERT
WIR MÜSSEN WISSEN
WIR WERDEN WISSEN
Hilbert's Epitaph
CANTOR
$2^{\aleph_0} = \aleph_1$
The Continuum Hypothesis

Marie Curie (1867-1934)
French of Polish descent

Nobel Prize in Physics, 1903
Nobel Prize in Chemistry, 1911
Marie Curie was an early pioneer in the study of radiation. She discovered the radioactive elements radium and polonium, the latter named after Poland, the country of her birth. Curie died at 66 from excessive exposure to radiation, the effects of which on human health were not at all understood at the time.

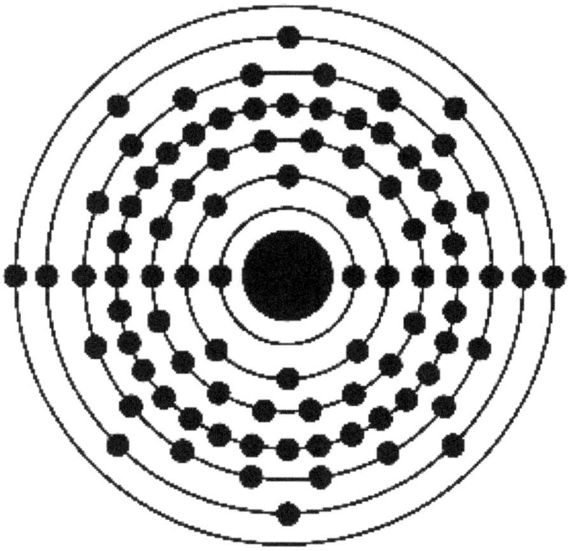

RADIUM

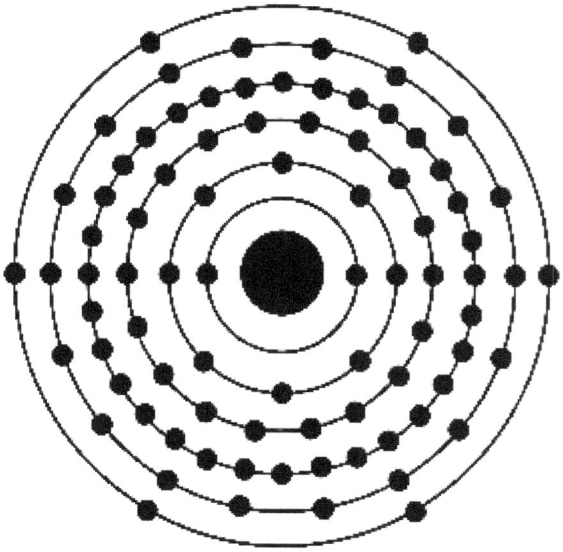

POLONIUM

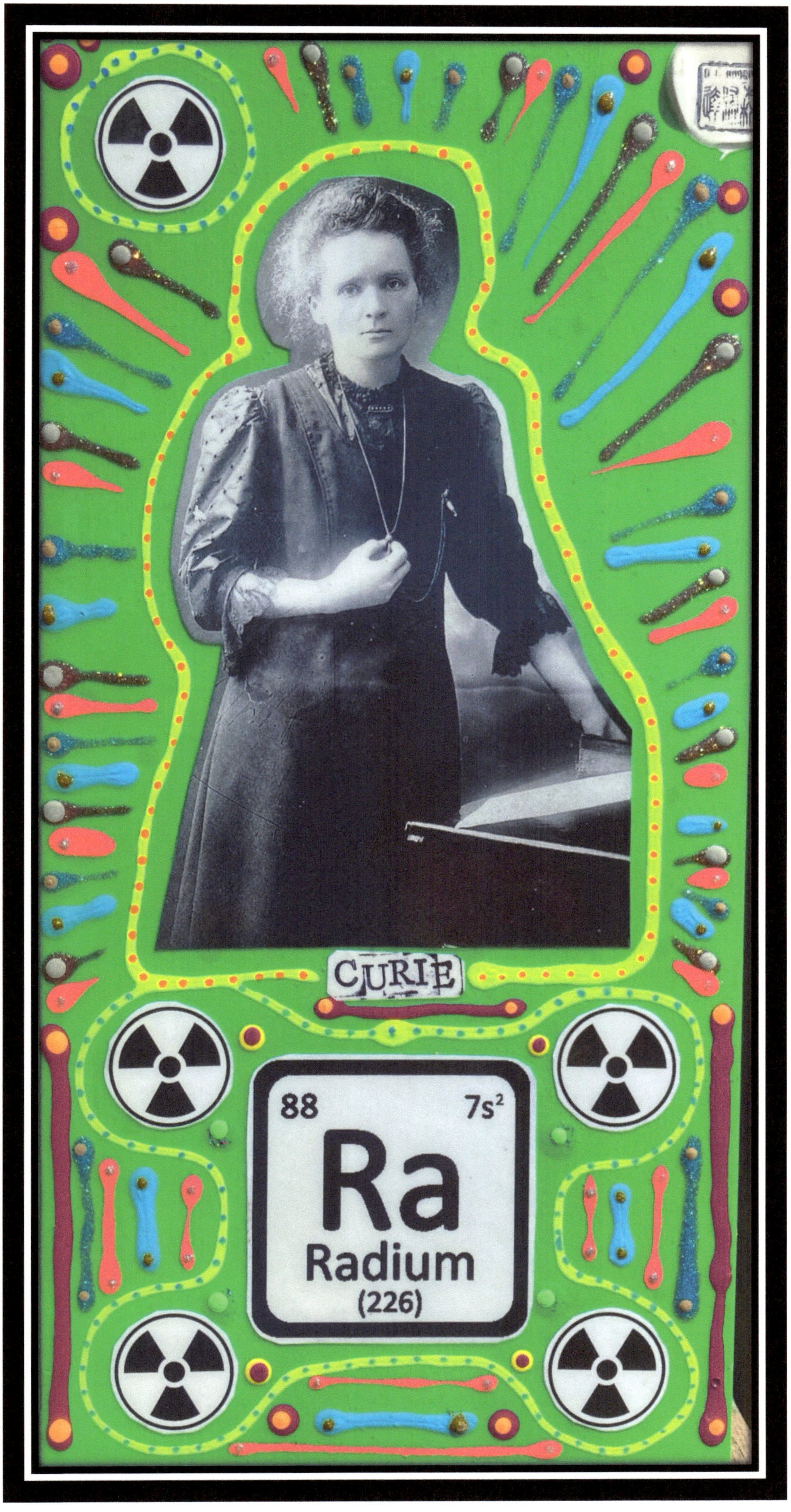
CURIE
88
7s²
Ra
Radium
(226)

Bertrand Russell (1872-1970), Alfred North Whitehead (1861-1947) and their *Principia Mathematica*

English

The Great Doomed Enterprise of math defined by math. It seemed like a good idea at the time. Russell was something of an intellectual celebrity during his life. He was a noted philosopher and was very outspoken about a lot of his beliefs. He was also a skilled symbolic logician. Whitehead was not well-known, but together, he and Russell composed the three-volume work that they called *Principia Mathematica*, or "Mathematical Principles." It is undoubtedly no coincidence that the title of Isaac Newton's 17th-century treatise on mathematical physics contains the same words.

The idea, which had percolated among mathematicians and logicians for some time, was to derive all mathematics from first principles using only inescapable symbolic logic. With *Principia Mathematica*, Russell and Whitehead believed they had done just that. The work, which is largely impenetrable to the amateur, was considered a success by those who could understand it. Then Gödel showed, some 20 years later, that any sufficiently sophisticated mathematics could contain unprovable, undecidable statements. This finding directly undercut the most basic assumption of Russell and Whitehead's project: that using pure logic, all mathematical statements could be proven either true or false.

Still, despite being basically discredited, Russell and Whitehead's *Principia Mathematica* remains an astonishing, monumental work of logic and a deep, deep dive into the fundamentals of mathematics.

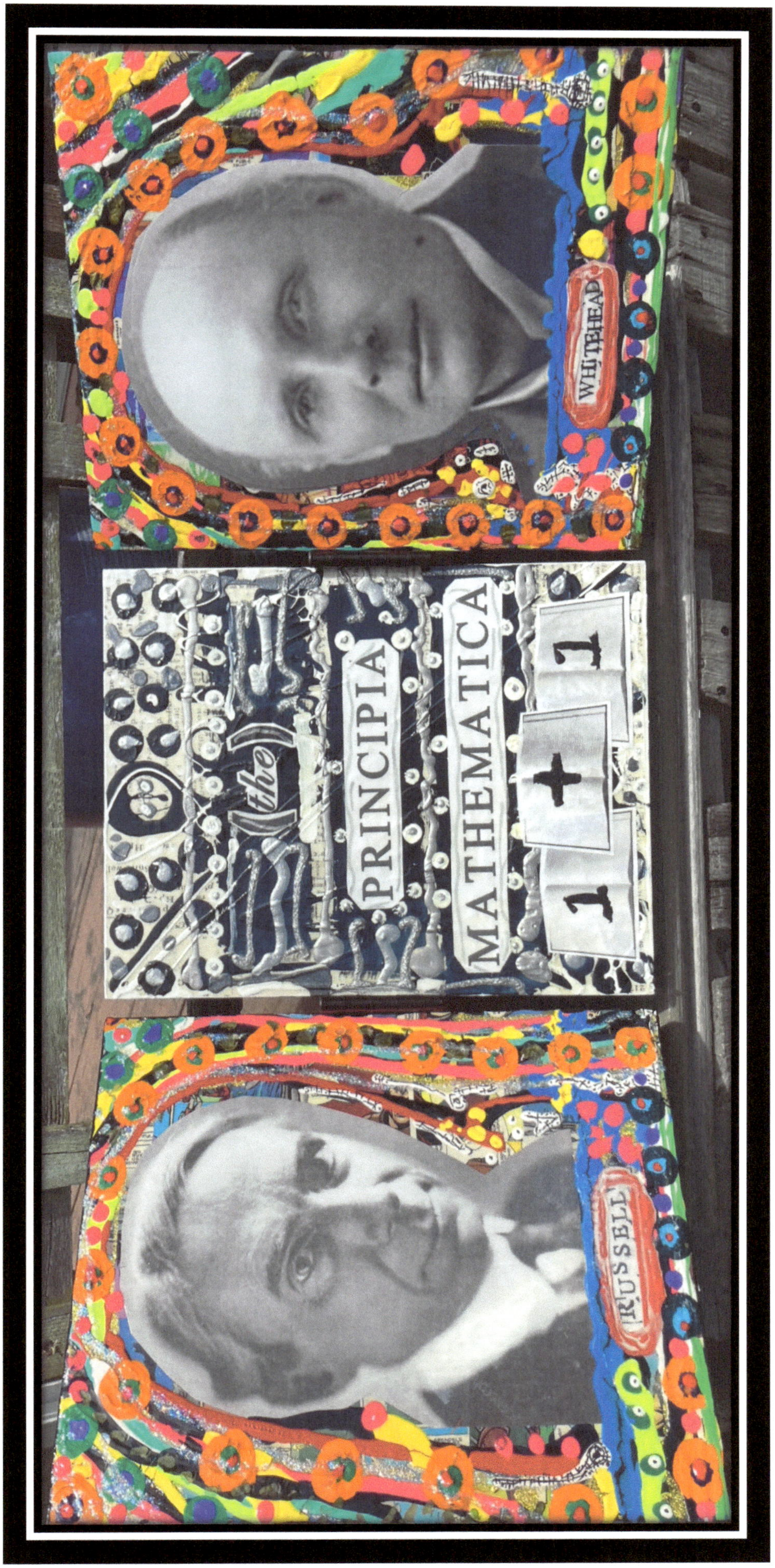
WHITEHEAD
the
PRINCIPIA
MATHEMATICA
1 + 1
RUSSELL

Emmy Noether (1882-1935)
German of Jewish descent

Noether's Theorem establishes a mathematical justification for conservation laws in physics. If you want to learn more, there are several good YouTube videos about it. Noether was primarily a mathematician and did a lot of important work on abstract algebra.

Noether's Theorem

$$\frac{d}{dt}\left(\sum_{\alpha} \frac{\partial \mathscr{L}}{\partial \dot{q}_{\alpha}} \xi_{\alpha}\right) = 0$$

Notice here that this equation is basically a derivative, a rate of change, that equals zero. Thus, it depicts an unchanging quantity.

Noether's Theorem says if a system has a continuous symmetry, then there must be corresponding quantities whose values are conserved.

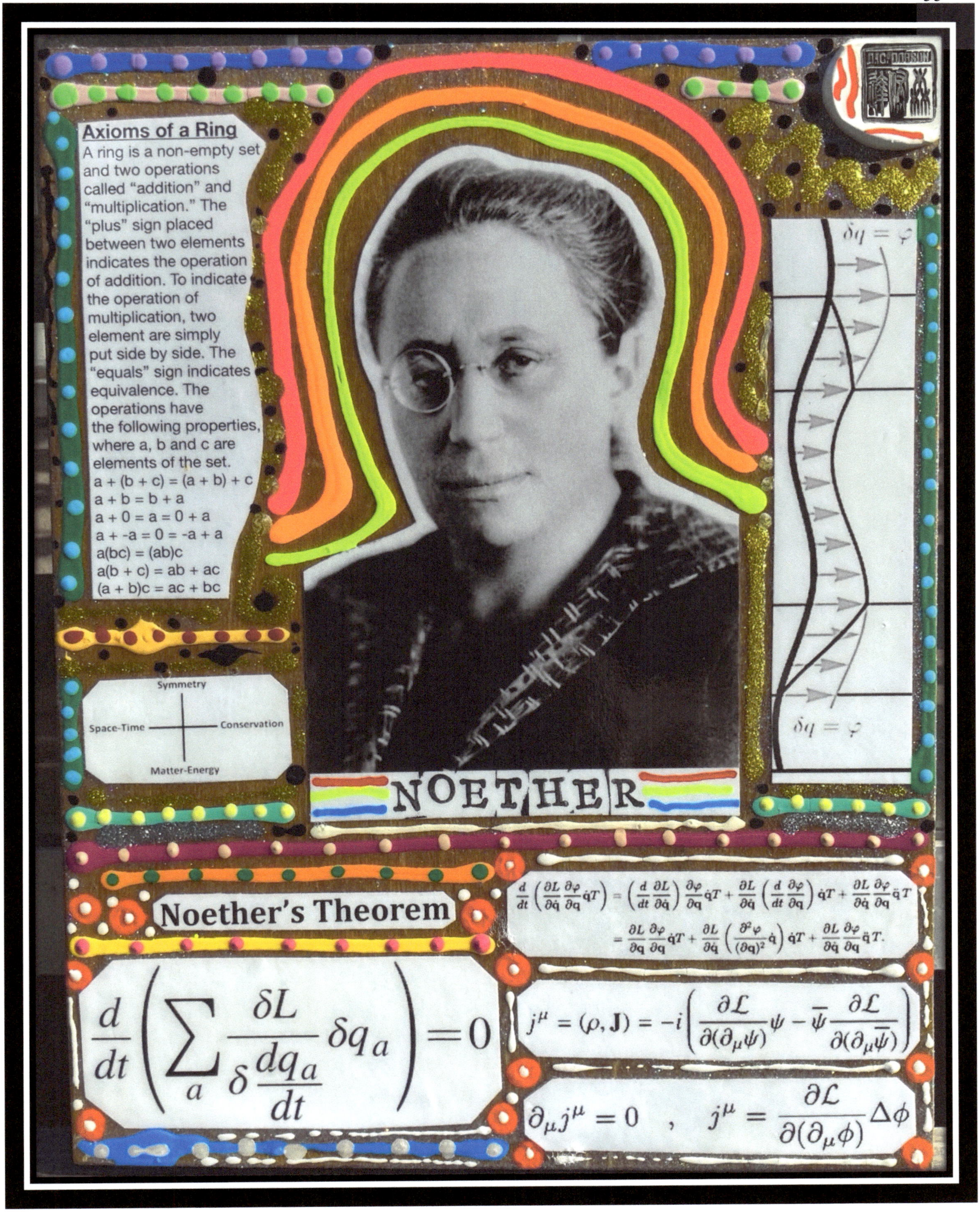
Axioms of a Ring
A ring is a non-empty set and two operations called "addition" and "multiplication." The "plus" sign placed between two elements indicates the operation of addition. To indicate the operation of multiplication, two element are simply put side by side. The "equals" sign indicates equivalence. The operations have the following properties, where a, b and c are elements of the set.
a + (b + c) = (a + b) + c
a + b = b + a
a + 0 = a = 0 + a
a + -a = 0 = -a + a
a(bc) = (ab)c
a(b + c) = ab + ac
(a + b)c = ac + bc
$\delta q = \varphi$
$\delta q = \varphi$
Symmetry
Space-Time
Conservation
Matter-Energy
NOETHER
Noether's Theorem
$\frac{d}{dt}\left(\sum_a \frac{\delta L}{\delta \frac{dq_a}{dt}} \delta q_a\right) = 0$
$\frac{d}{dt}\left(\frac{\partial L}{\partial \dot{\mathbf{q}}}\frac{\partial \varphi}{\partial \mathbf{q}}\dot{\mathbf{q}}T\right) = \left(\frac{d}{dt}\frac{\partial L}{\partial \dot{\mathbf{q}}}\right)\frac{\partial \varphi}{\partial \mathbf{q}}\dot{\mathbf{q}}T + \frac{\partial L}{\partial \dot{\mathbf{q}}}\left(\frac{d}{dt}\frac{\partial \varphi}{\partial \mathbf{q}}\right)\dot{\mathbf{q}}T + \frac{\partial L}{\partial \dot{\mathbf{q}}}\frac{\partial \varphi}{\partial \mathbf{q}}\ddot{\mathbf{q}}T$
$= \frac{\partial L}{\partial \mathbf{q}}\frac{\partial \varphi}{\partial \mathbf{q}}\dot{\mathbf{q}}T + \frac{\partial L}{\partial \dot{\mathbf{q}}}\left(\frac{\partial^2 \varphi}{(\partial \mathbf{q})^2}\dot{\mathbf{q}}\right)\dot{\mathbf{q}}T + \frac{\partial L}{\partial \dot{\mathbf{q}}}\frac{\partial \varphi}{\partial \mathbf{q}}\ddot{\mathbf{q}}T.$
$j^\mu = (\rho, \mathbf{J}) = -i\left(\frac{\partial \mathcal{L}}{\partial(\partial_\mu \psi)}\psi - \bar{\psi}\frac{\partial \mathcal{L}}{\partial(\partial_\mu \bar{\psi})}\right)$
$\partial_\mu j^\mu = 0 \quad , \quad j^\mu = \frac{\partial \mathcal{L}}{\partial(\partial_\mu \phi)}\Delta\phi$

Albert Einstein (1879-1955), Richard Feynman (1918-1988) and Robert Oppenheimer (1904-1967)

German American (Einstein) and
American (Feynman and Oppenheimer) of Jewish descent

Relativity, quantum electrodynamics and the bomb. Einstein was born in Germany but became an American citizen later in his life. Feynman and Oppenheimer were both born in America.

Jews were not allowed (by their own religious authorities) to study secular subjects such as math and physics until about 1850, a change contemporaneous with the advent of Reform Judaism, and indeed with a Judaism-wide attempt to reconcile their ancient religious laws with the rapidly encroaching modern world. Gentile antagonism in the form of pogroms also contributed to the desire of many Jews to seek a new way. The modern Haredi (aka ultra-Orthodox) movement emerged at around the same time, as did modern Zionism. Since then, many Jews have excelled in the sciences.

Einstein's theories of Special and General Relativity, published in 1905 and 1916, respectively, continue to be the best available description of gravity and light. Contrary to popular belief, Einstein did not fail high school math. He actually did very well. Einstein received the Nobel Prize in Physics in 1921.

Feynman was known for his eccentric, charismatic personality. His lectures could be called flamboyant, but really he just had a passion for physics. Feynman Diagrams were very important in the development of quantum electrodynamics, the quantum theory of the electron and electromagnetism. Feynman received the Nobel Prize in Physics in 1965.

Oppenheimer, as everyone knows, led the Manhattan Project, the American effort to develop atomic weapons during World War II. He succeeded with a bang!

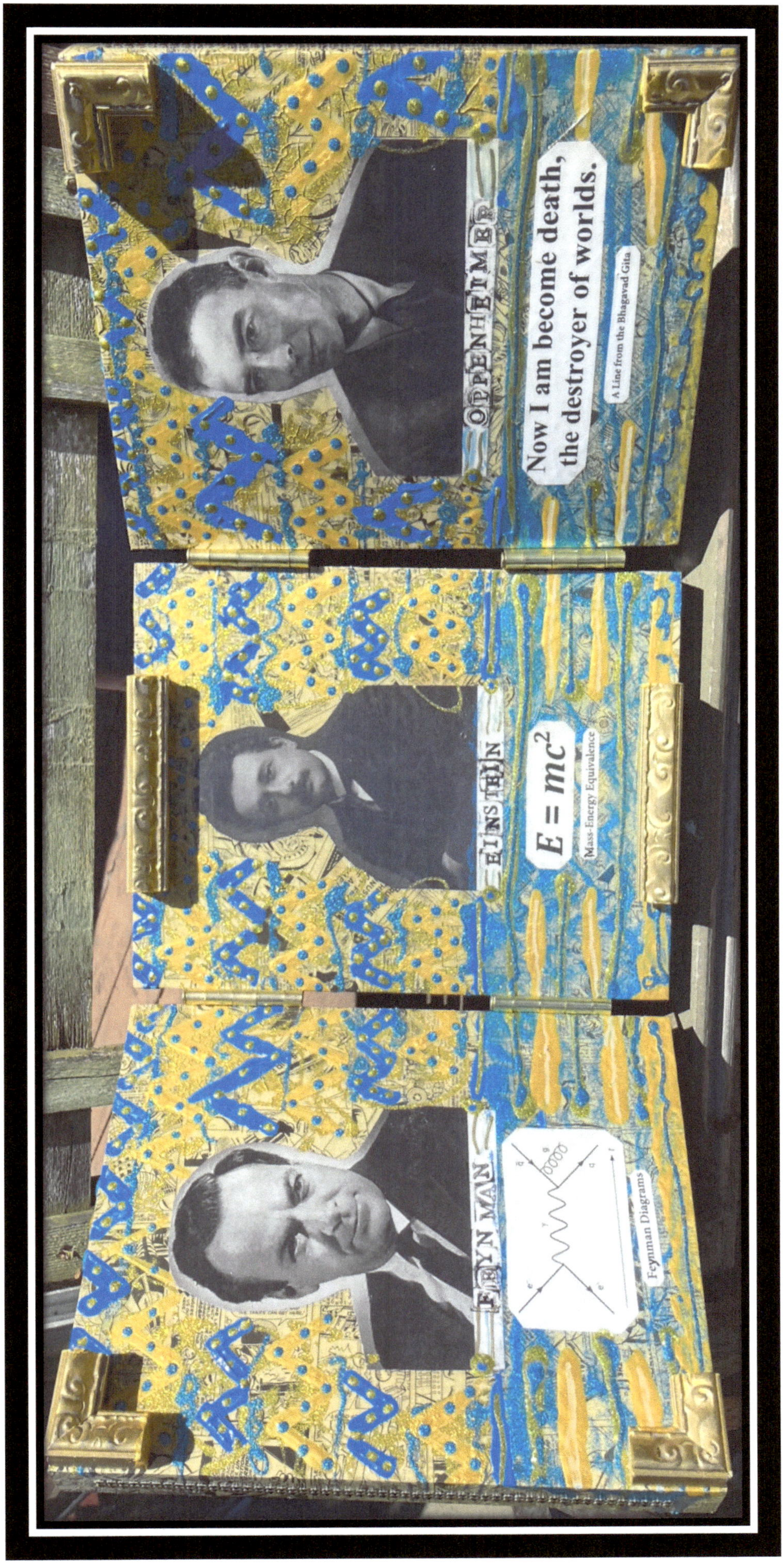
OPPENHEIMER
Now I am become death, the destroyer of worlds.
A Line from the Bhagavad Gita
EINSTEIN
E = mc2
Mass-Energy Equivalence
FEYNMAN
Feynman Diagrams

Max Born (1882-1970)
German of Jewish descent

Max Born made innumerable contributions to physics in the age of quantum theory. He was awarded the Nobel Prize in Physics in 1954. The Born Rule, shown here, says that the probability of finding a particle at a certain location is equal to the square of the wavefunction at that location.

Born approximation
Born coordinates
Born equation
Born probability
Born reciprocity
Born rigidity
Born series
Born square
Born–Landé equation
Born–Infeld theory
Born–Haber cycle
Born–Huang approximation
Born–von Karman boundary condition
Born–Oppenheimer approximation
BBGKY hierarchy
Cauchy–Born rule
Adiabatic theorem
Canonical commutation relation

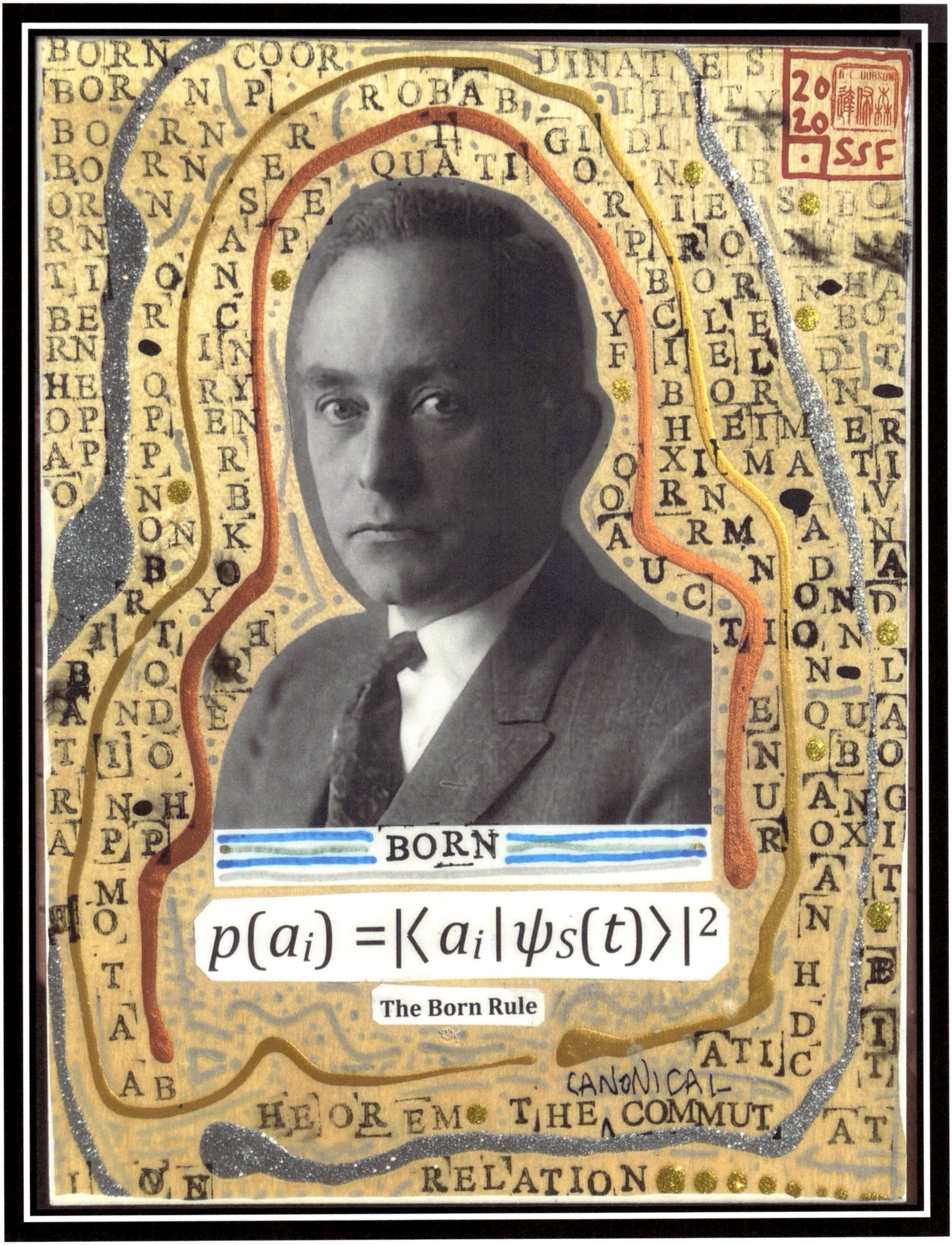

2020 SSF
BORN
$p(a_i) = |\langle a_i | \psi_S(t) \rangle|^2$
The Born Rule
CANONICAL
THE COMMUT
RELATION

Niels Bohr (1885-1962) and Werner Heisenberg (1901-1976)
Danish (Bohr) and German (Heisenberg)

Two early quantum theorists. Bohr is known for his quantum model of the atom, wherein electrons make discrete jumps between energy levels. Heisenberg is known for his Uncertainty Principle, which states that we cannot know a particle's exact position and exact momentum at the same time, as the product of change in position and change in momentum is always greater than zero.

The Bohr Model of the Atom

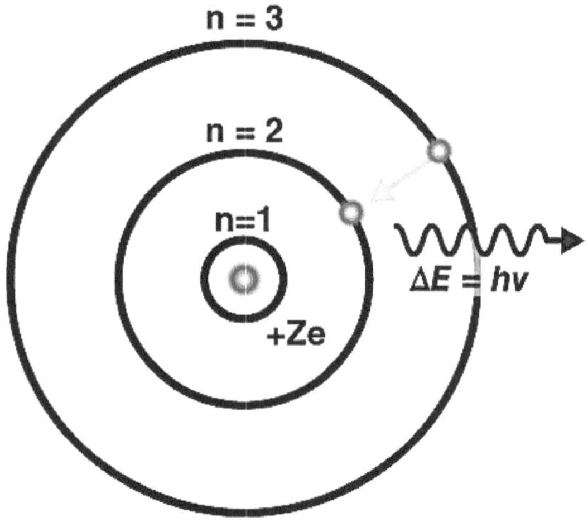

Notice here that electron orbits are discrete jumps from one level to another.

The Heisenberg Uncertainty Principle

$$\Delta\chi\Delta\rho \geq \frac{\hbar}{2}$$

Notice here that change in position times change in momentum is always greater than zero. So neither can ever equal zero. For subatomic particles, position and momentum are always changing and moving.

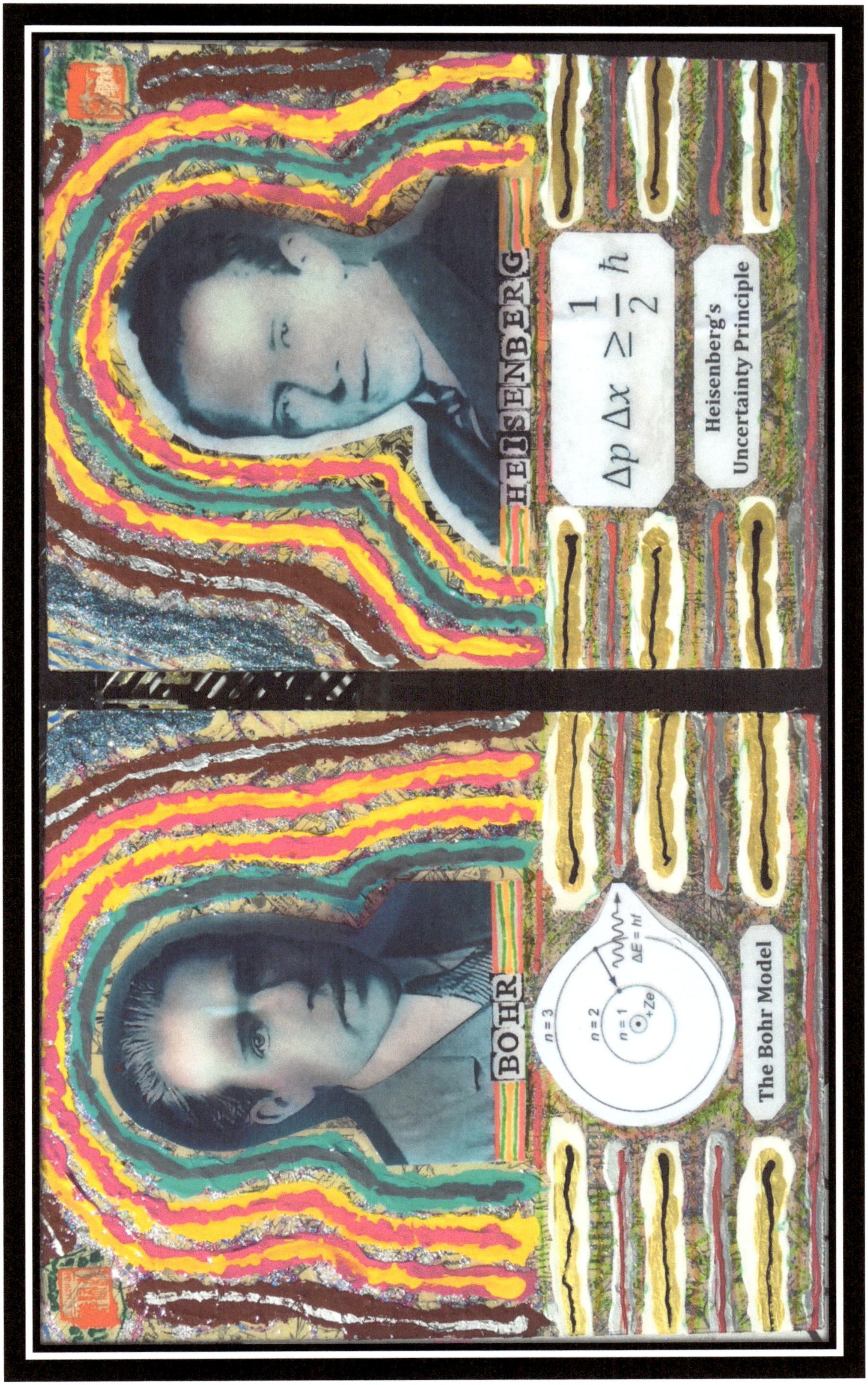
BOHR
n = 3
n = 2
n = 1
+Ze
ΔE = hf
The Bohr Model
HEISENBERG
$\Delta p \, \Delta x \geq \frac{1}{2} h$
Heisenberg's
Uncertainty Principle

Paul A. M. Dirac (1902-1984) and
Erwin Schrodinger (1887-1961)
English (Dirac) and German (Schrodinger)

Dirac and Schrodinger were jointly awarded the Nobel Prize in physics in 1933. Among other achievements, Dirac invented the field of quantum electrodynamics, which is the quantum theory of the electron, and thus of electromagnetism and of light. Schrodinger is known for formulating the quantum wavefunction and introducing the notion of superposition. Schrodinger is also responsible for the question of his cat, but that tale is well known.

Different Forms of Schrodinger's equation

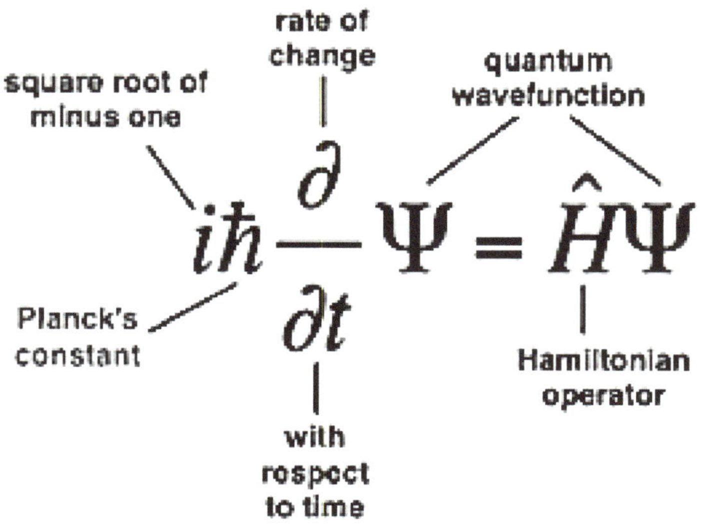

$$H\Psi = i\hbar\frac{\partial\Psi}{\partial t}$$

Time evolution

$$\frac{-\hbar^2}{2m}\frac{\partial^2\Psi(x)}{\partial x^2} + U(x)\Psi(x) = E\Psi(x)$$

Time independent equation

Different Form of Dirac's Equation

$$(\mathrm{i}\hbar\gamma^{\mu}\nabla_{\mu} - \mathrm{mc})\psi = 0$$

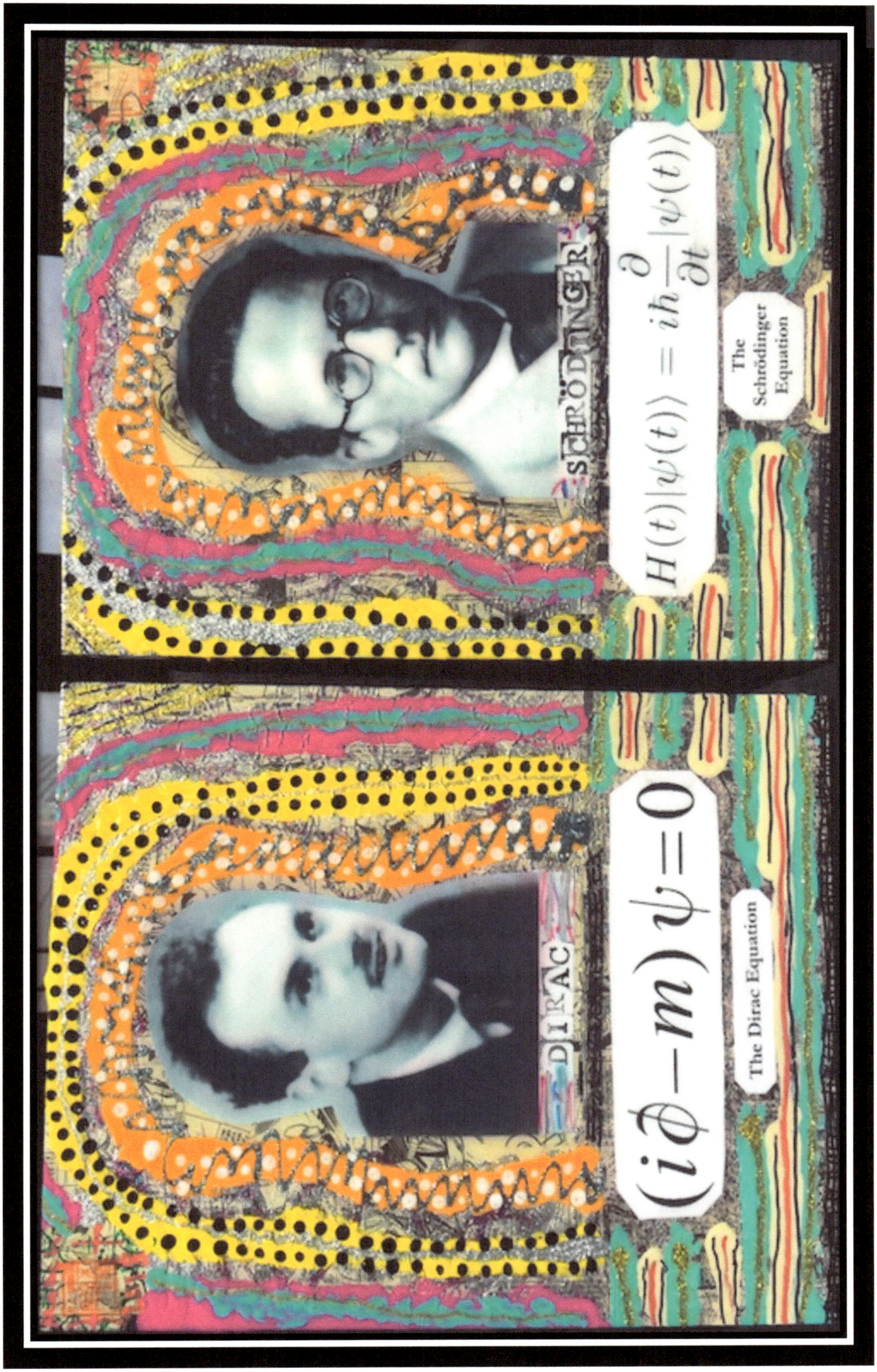
DIRAC
$(i\partial\!\!\!/ - m)\psi = 0$
The Dirac Equation
SCHRÖDINGER
$H(t)|\psi(t)\rangle = i\hbar\frac{\partial}{\partial t}|\psi(t)\rangle$
The Schrödinger Equation

Louis De Broglie (1897-1987)
French

De Broglie ("de BROHG-lee"), a French aristocrat, was an important contributor to early quantum physics. He firmly established the dual, wave-particle nature of matter, especially at extremely small scales. In this painting, we see his simple equation that relates wavelength to mass using Planck's constant. De Broglie received the Nobel Prize in Physics in 1929.

De Broglie's Model of the Atom
In which Electrons are Standing Waves

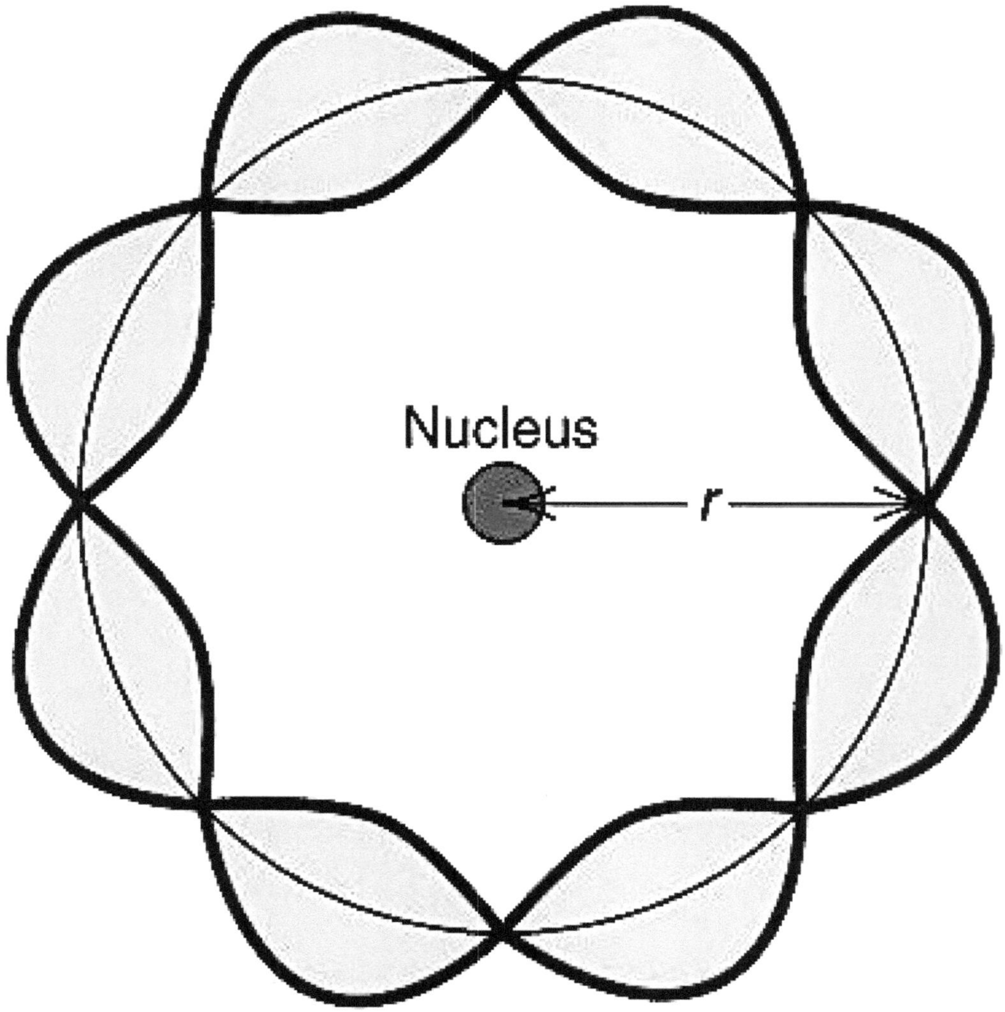

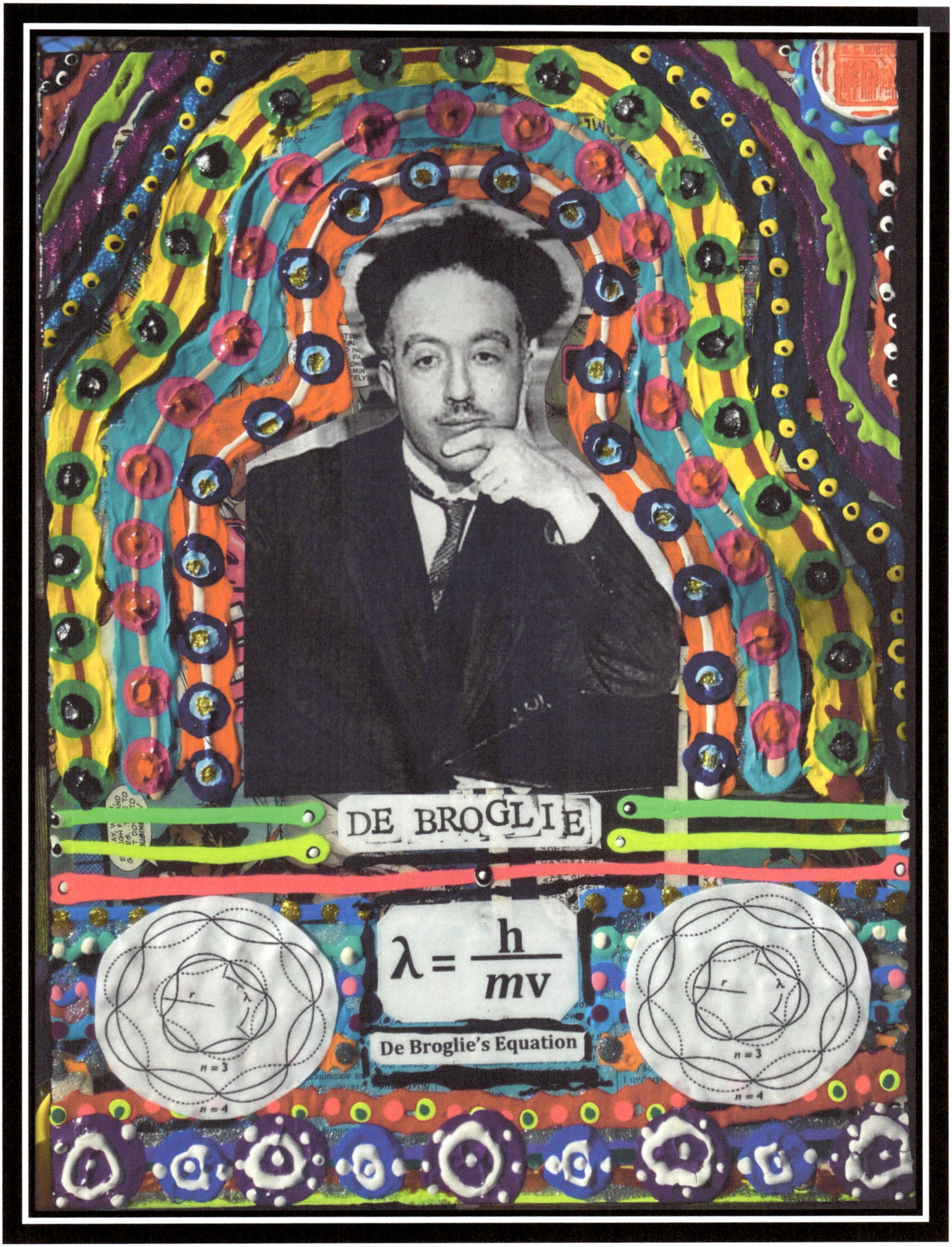
DE BROGLIE
$\lambda = \frac{h}{mv}$
De Broglie's Equation
r
λ
n = 3
n = 4
r
λ
n = 3
n = 4

Wolfgang Pauli (1900-1958)

Austrian

Nobel Prize in Physics, 1945

Pauli was a theoretical physicist in the early days of quantum physics. In this painting, we see three of his main achievements.

Pauli was one of the first to theorize the existence of neutrinos, an electrically neutral subatomic particle. The existence of neutrinos is an important part of particle physics. Neutrinos are hard to detect since they do not interact much with ordinary matter, but their existence was first confirmed experimentally in 1956. It is well understood that trillions of neutrinos, mostly generated by the sun, pass through your body every day.

The Pauli Exclusion Principle says that no two electrons can simultaneously occupy the same quantum state. The quantum state is defined by four quantum numbers, one of which is spin, the intrinsic angular momentum of the particle.

Pauli pioneered the use of what he called "regularization," known today as renormalization. In the mathematics of quantum physics, infinities can appear in places where they render the theory nonsensical. Regularization is a way of sweeping inconvenient infinities under the rug.

Pauli was born in Vienna to a Catholic mother and Jewish father. He was raised Catholic. His godfather was the Austrian physicist Ernst Mach, whose name is attached to the speed of sound. The young Pauli was a distinguished student. He obtained his PhD from the University of Munich in 1921. Soon after, he wrote a seminal paper on Einstein's relativity. He held positions at Göttingen, Copenhagen, Hamburg, Zürich and at the IAS in New Jersey. Pauli was known as the consummate theoretician. It was said that whenever he entered the room, the experimental apparatus would break down.

Pauli also had a close and productive relationship with the Swiss psychologist Carl Jung, who analyzed many of Pauli's dreams.

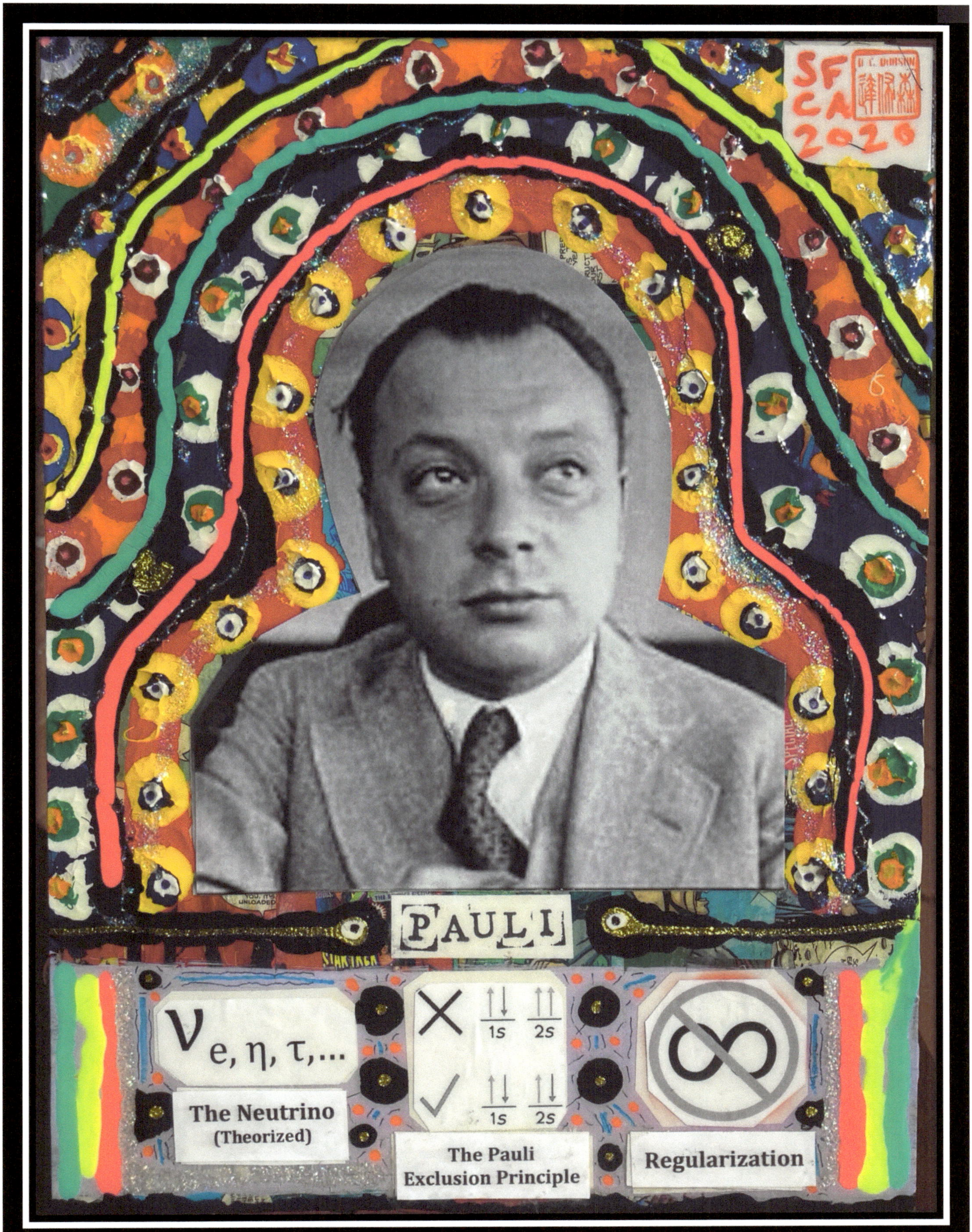
SF
CA
2020
PAULI
$\nu_{e, \eta, \tau, \ldots}$
The Neutrino
(Theorized)
1s 2s
1s 2s
The Pauli
Exclusion Principle
Regularization

Edwin Hubble (1889-1953)
American of mixed European descent

Hubble benefitted from the advances in telescope technology that occurred in his lifetime. With new telescopes, Hubble was able to see much farther and more clearly into the night sky than his predecessors. In particular, he was able to determine the distances to other stars and galaxies.

Hubble's observations determined that the Milky Way is not the only galaxy, and that there are galaxies outside it. He also discovered the red shift of far distant galaxies and that they all seem to be moving away from the Milky Way in all directions. This led to the belief that the universe is expanding, but not with Earth at its center. Rather, it is thought that at a very large scale, everything is moving away from everything else, such that, if we were transported to another place in the universe, we would observe the same phenomenon.

It's not a perfect analogy but imagine drawing dots on a rubber balloon and then inflating it. All the dots would move farther and farther from one another because the very space between them is growing. Of course, on the balloon, the dots too would grow, but this does not seem to be the case for galaxies and clusters of galaxies, all of which are bound together by gravity in one way or another.

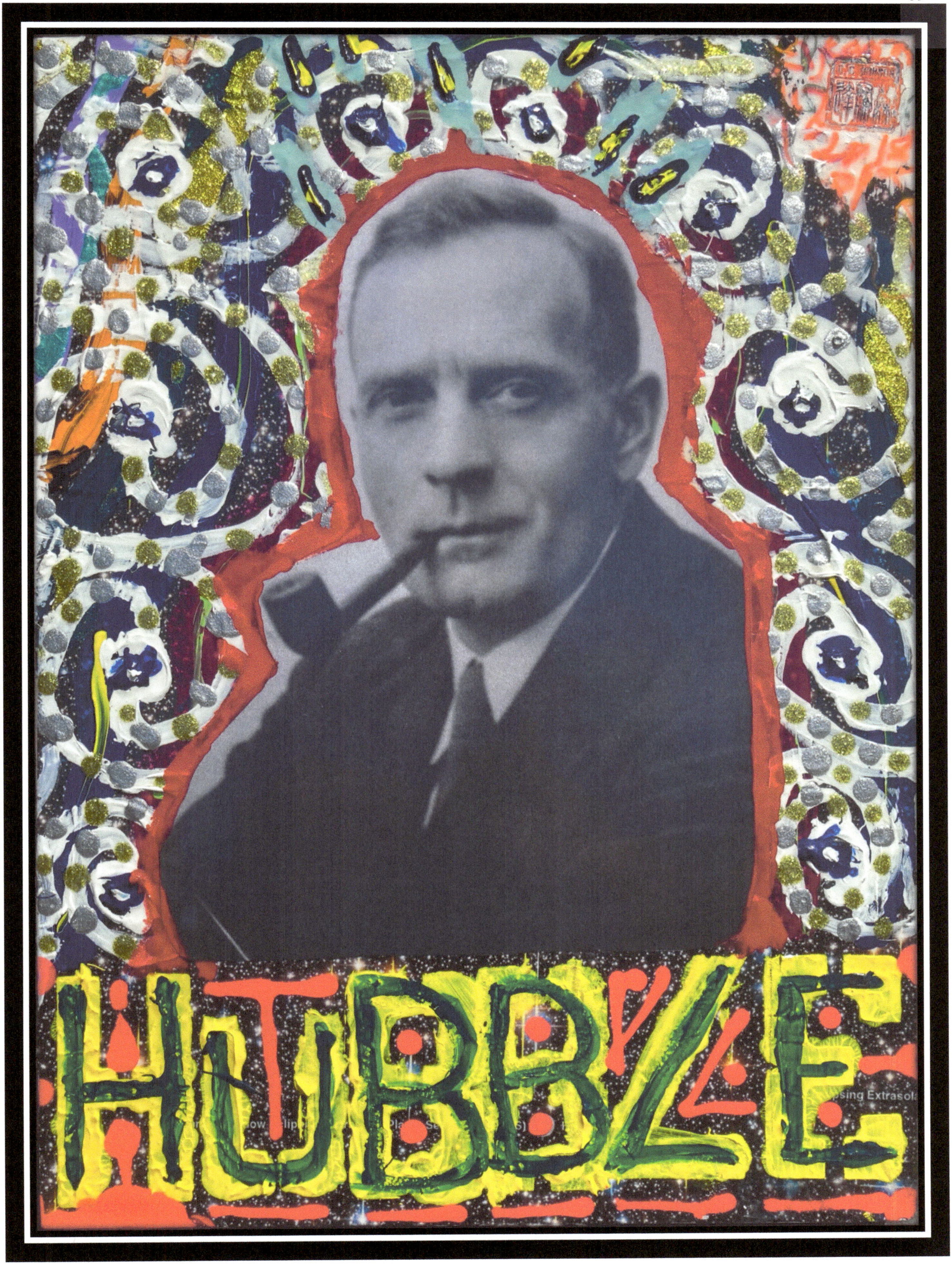
HUBBLE

Srinivasa Ramanujan (1887-1920) and Subramanyan Chandrasekhar (1910-1995)

Indian (Ramanujan) and Indian American (Chandrasekhar)

Two 20th-century Indian scientists. Ramanujan was an almost entirely self-taught mathematical savant and prodigy from Mysore, India, who ultimately went to study in England.

"In 1913, Ramanujan, seeking mathematicians who could better understand his work, began a mail correspondence with the English mathematician G. H. Hardy at the University of Cambridge, England. Recognizing Ramanujan's work as extraordinary, Hardy arranged for him to travel to Cambridge. In his notes, Hardy commented that Ramanujan had produced groundbreaking new theorems, including some that 'defeated me completely; I had never seen anything in the least like them before.'"
Wikipedia, "Srinivasa Ramanujan", Tuesday June 18, 2024.

Unfortunately, Ramanujan passed away very young, at the age of 32, from tuberculosis and malnutrition. He died in India in 1920.

Chandrasekhar's Limit is the minimum amount of mass a star must possess in order to collapse into a black hole. Any star more massive than this limit can potentially collapse into a black hole at the end of its life. A black hole is what happens when a star's gravity, due to its great mass, completely overwhelms any outward pressure from within and collapses in on itself down to that inscrutable and vexing point known as a singularity.

Chandrasekhar, who was born in Lahore before the 1947 partition of British India, had a long and distinguished career as a physicist and astrophysicist. He worked for the war effort in America during World War II. He stayed in the US, and he and his wife became American citizens in 1953.

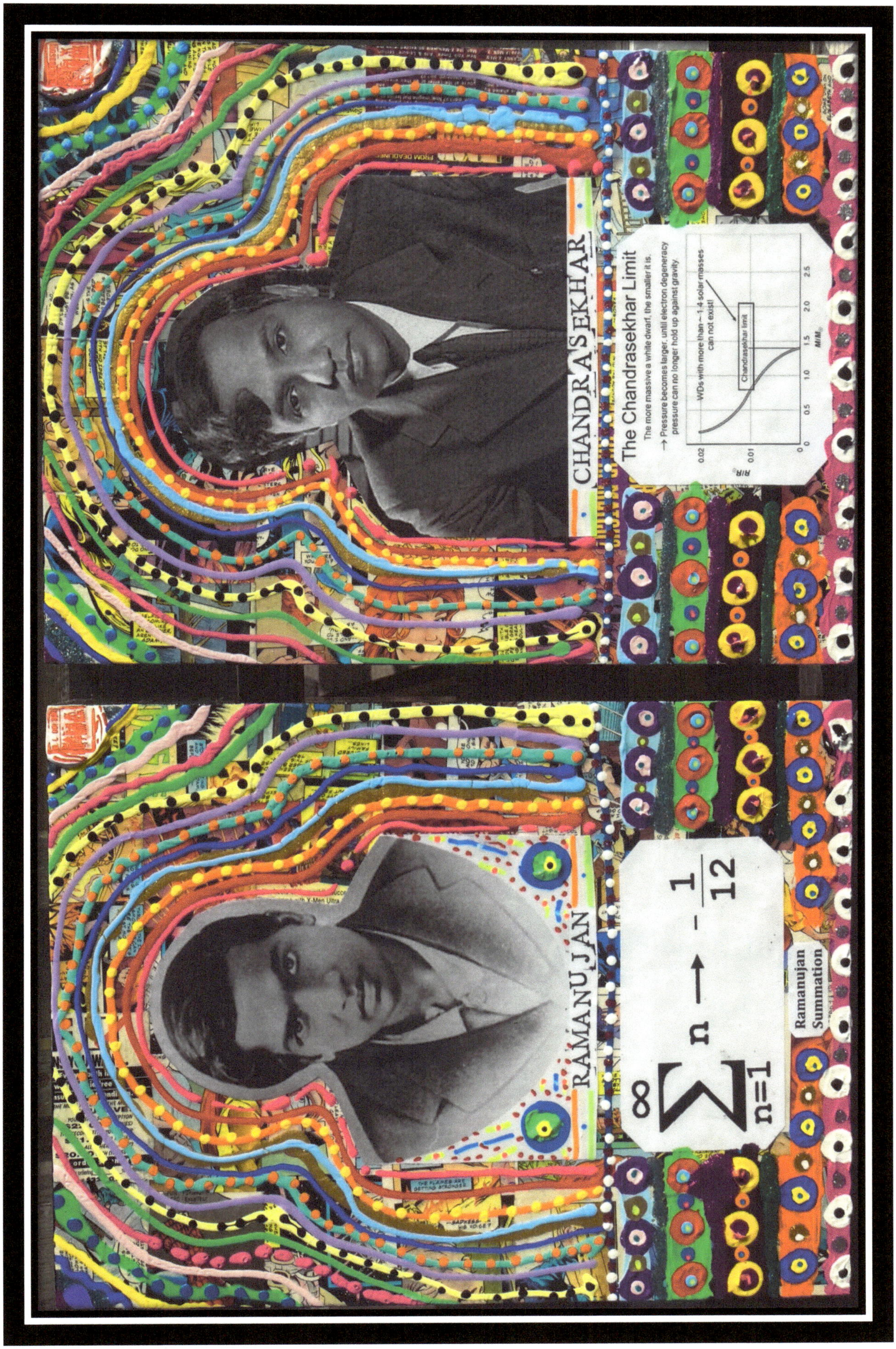
CHANDRASEKHAR
The Chandrasekhar Limit
The more massive a white dwarf, the smaller it is.
→ Pressure becomes larger, until electron degeneracy pressure can no longer hold up against gravity.
WDs with more than ~1.4 solar masses can not exist!
Chandrasekhar limit
RAMANUJAN
$\sum_{n=1}^{\infty} n \longrightarrow -\frac{1}{12}$
Ramanujan Summation

Kurt Gödel (1906-1978) and Alan Turing (1912-1954)
Austrian (Gödel) and English (Turing)

With his Incompleteness Theorems, Gödel showed that any sufficiently sophisticated mathematics can produce unprovable, undecidable statements. Turing proved that his infamous Halting Problem cannot be decided. So, both proved a negative, Turing on the heels of Gödel.

A Turing Machine

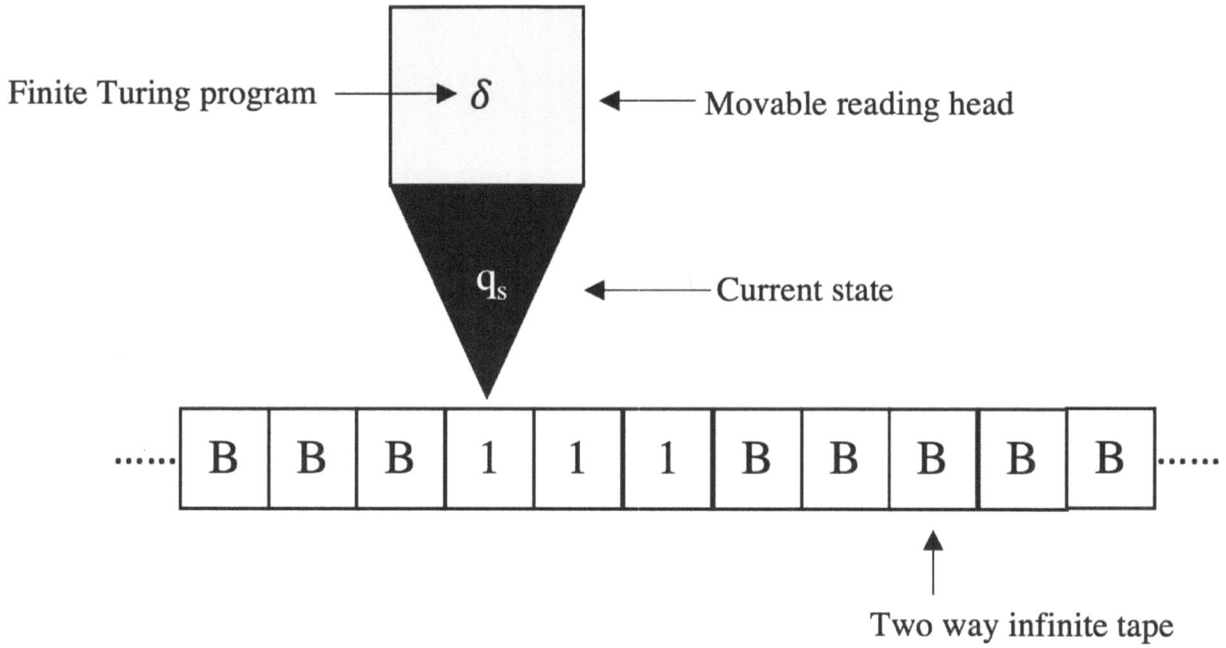

A Turing Machine is an imaginary machine that would in theory be able to carry out any calculations that a computer could.

Godel's First Incompleteness Theorem

$$G \leftrightarrow \neg \mathrm{Prov}(\ulcorner G \urcorner)$$

"G if and only if cannot prove G"

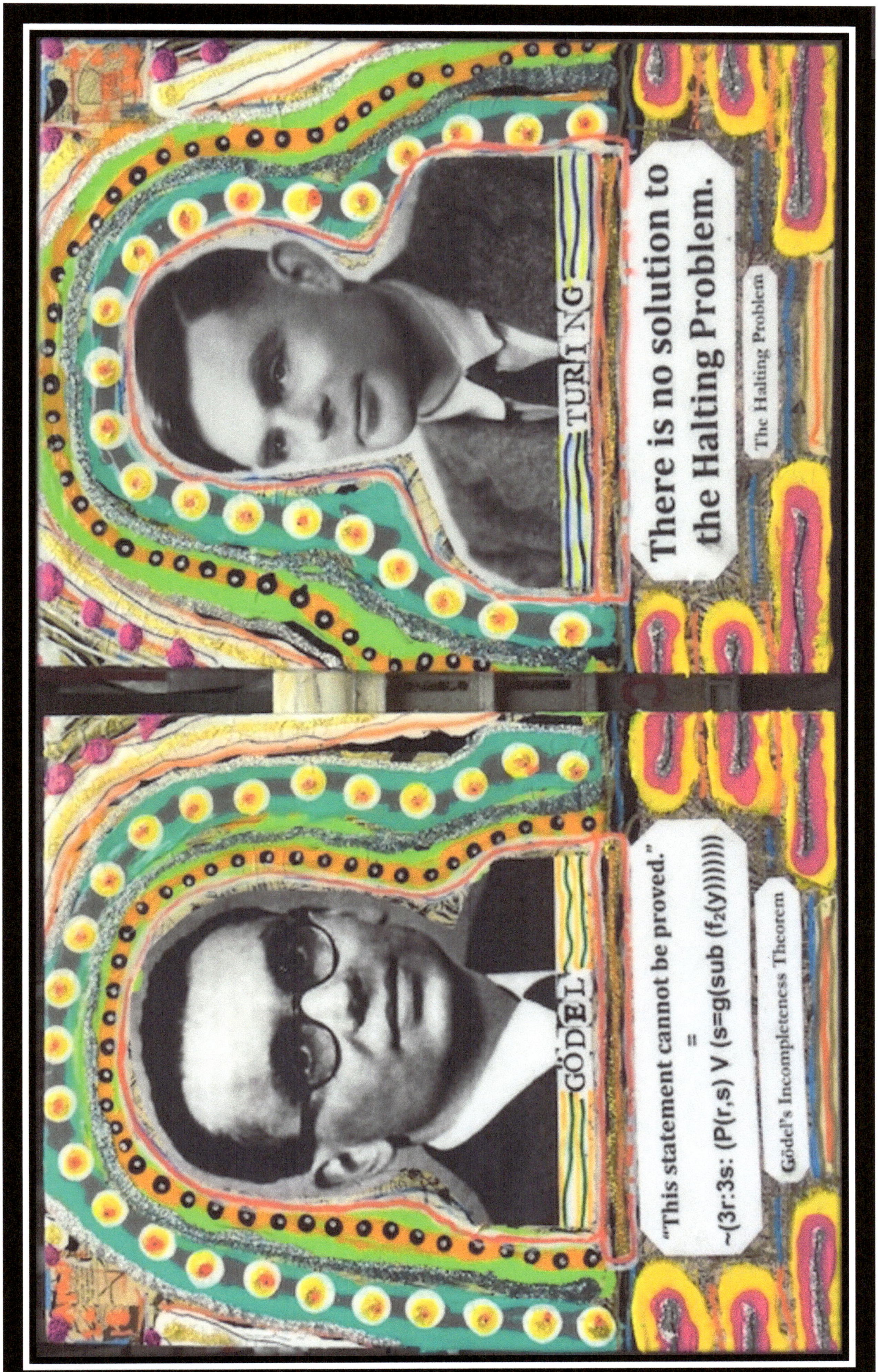
GÖDEL
"This statement cannot be proved."
=
~(∃r:∃s: (P(r,s) V (s=g(sub (f2(y))))))
Gödel's Incompleteness Theorem
TURING
There is no solution to the Halting Problem.
The Halting Problem

Enrico Fermi (1901-1954)
American of Italian descent

Fermi, born in Rome, Italy, oversaw the building of the first manmade nuclear reactor in Chicago, Illinois. It was a huge pile of uranium and graphite.

"Fermi left Italy in 1938 to escape new Italian racial laws that affected his Jewish wife, Laura Capon. He emigrated to the United States, where he worked on the Manhattan Project during World War II. Fermi led the team at the University of Chicago that designed and built Chicago Pile-1, which went critical on 2 December 1942, demonstrating the first human-created, self-sustaining nuclear chain reaction."
Wikipedia, "Enrico Fermi," Wednesday, June 19, 2024.

Fermi became a naturalized American citizen in 1944. He died of stomach cancer, which may have been caused by excessive exposure to radiation.

Nuclear Fission

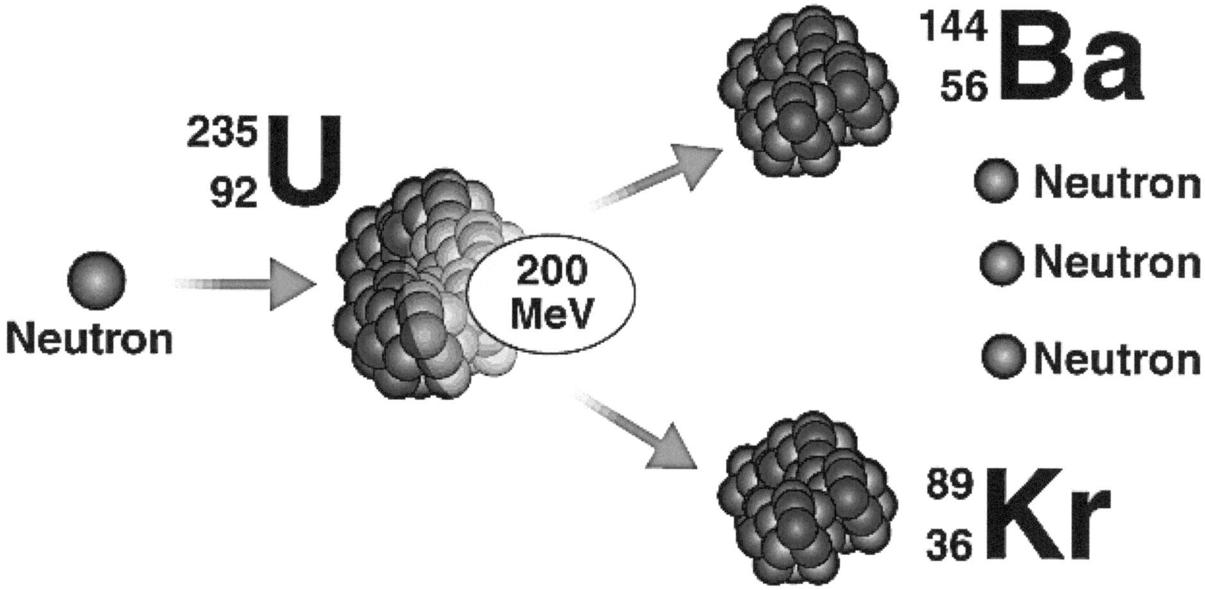

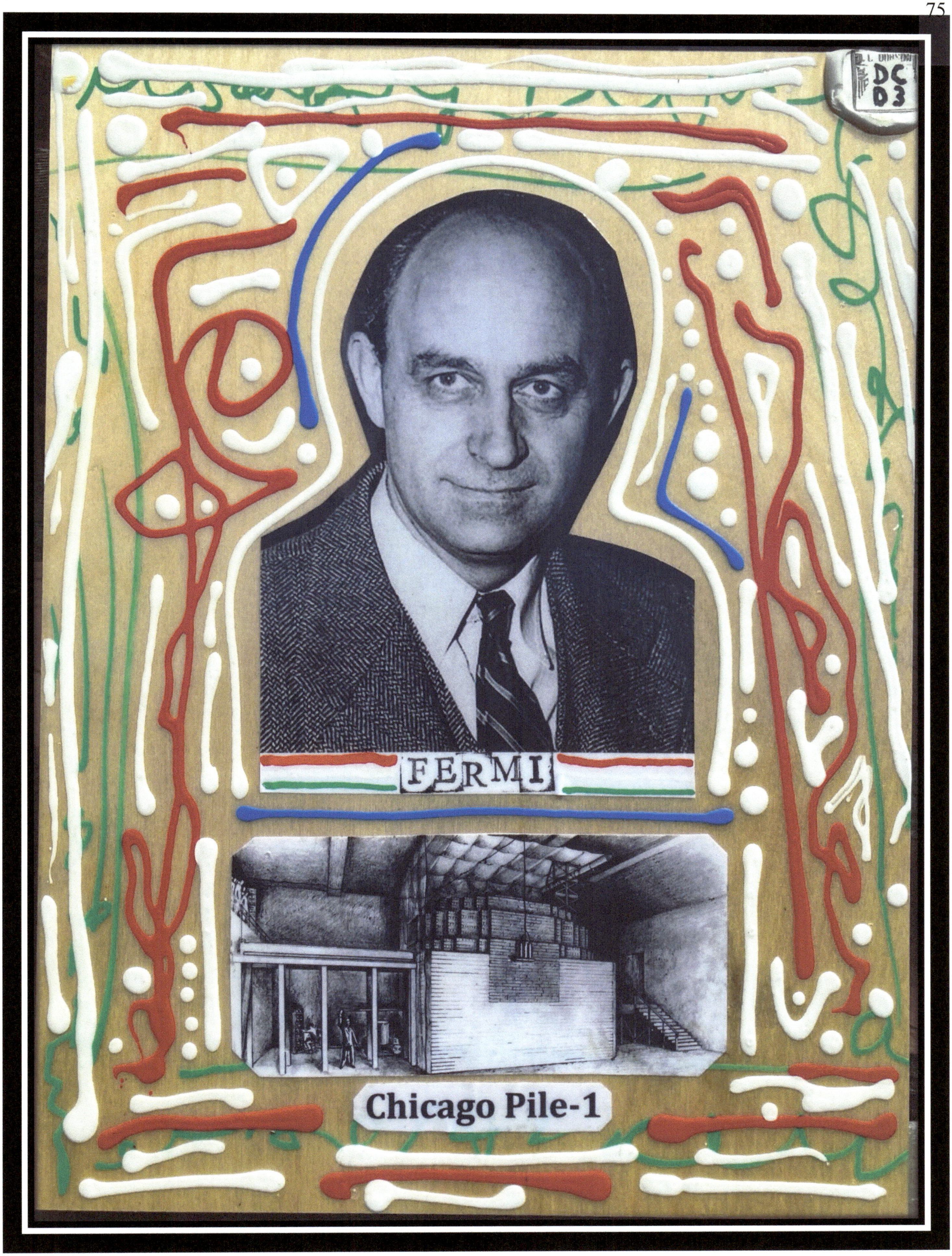
DC
D3
FERMI
Chicago Pile-1

Paul Erdos (1913-1996)

Hungarian of Jewish descent

Erdos is known as one of the most prolific mathematicians in history. He was author or co-author of around 1500 papers. Only Euler wrote more pages, but Erdos published more separate papers. Also, in Erdos's case, the vast majority of his papers are collaborations with one or more other authors, while Euler is the sole author of his works. In this painting we see some of the titles of his many collaborations.

It seems that Erdos was an itinerant mathematician. I think he slept on many a couch. He would basically stay with a particular mathematician for a couple weeks, churn out a paper and then move on. He would also attend conferences and seminars. Erdos saw mathematics as a social activity, one meant to be engaged in with other people. This flies in the face of the traditional model of the heroic lone mathematician working in isolation. Erdos and his companions solved numerous outstanding problems and also posed many new, unsolved ones.

There is something among mathematicians called an Erdos Number. Erdos himself has the number 0 (zero). People who collaborated directly with Erdos have Erdos Number 1 (one). If you collaborated with someone with EN 1, then you have EN 2, and so on.

Born in Budapest to two high school math teachers, Erdos learned much about mathematics as a young child. He earned his PhD in mathematics from the University of Budapest at age 21. He moved to the United States in 1938 and spent much of his life travelling around the country, holding positions at the IAS, the University of Pennsylvania, Notre Dame, Purdue and Stanford, among others. Erdos never married and had no children. He died of a heart attack while attending a mathematical conference in Warsaw, Poland.

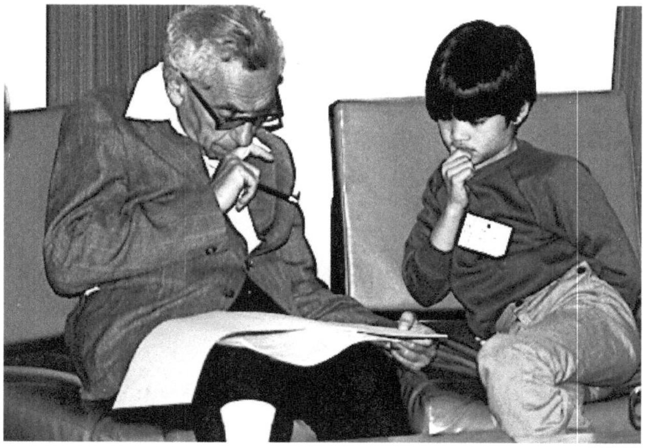

Paul Erdos and Terry Tao in 1985

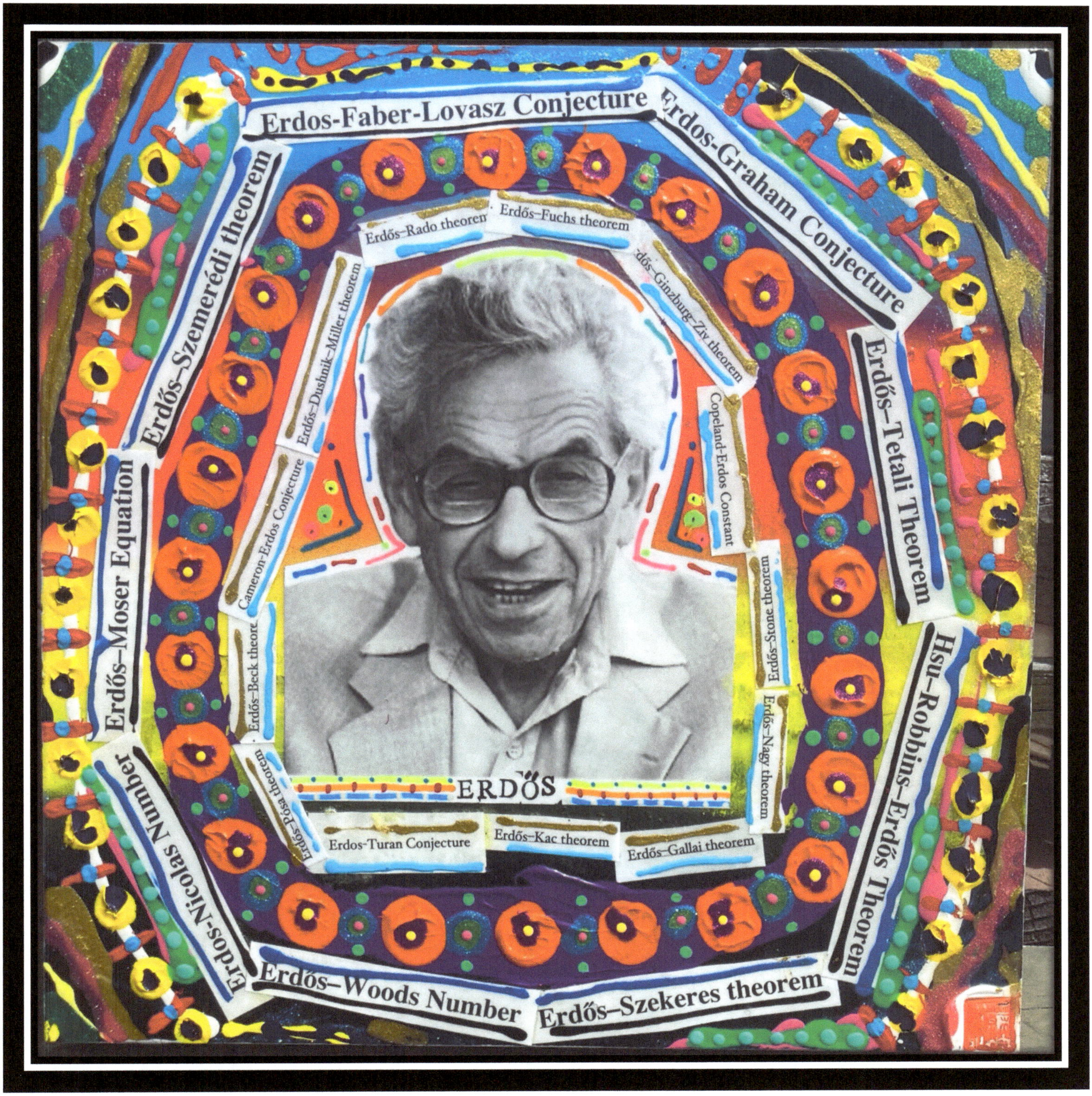
Erdos-Faber-Lovasz Conjecture
Erdos-Graham Conjecture
Erdős–Szemerédi theorem
Erdős–Rado theorem
Erdős–Fuchs theorem
dős–Ginzburg–Ziv theorem
Erdős–Dushnik–Miller theorem
Copeland-Erdos Constant
Erdős–Tetali Theorem
Erdős–Moser Equation
Cameron-Erdos Conjecture
Erdős–Stone theorem
Erdős–Beck theore
Erdős–Nagy theorem
Hsu-Robbins-Erdős Theorem
ERDŐS
Erdős–Pósa theorem
Erdos-Turan Conjecture
Erdős–Kac theorem
Erdős–Gallai theorem
Erdos-Nicolas Number
Erdős–Woods Number
Erdős–Szekeres theorem

Alexander Grothendieck (1928-2014)
French of German descent

Algebraic geometry. Topology. Number theory. Sheaves. Fields Medal 1966. I don't get it. Way over my head. For more information on the Fields Medal, please consult Appendix J.

Grothendieck Construction

$$\int_{c\in\mathscr{C}} F(c) = \left\{ \begin{array}{c} * \\ \overset{a}{\swarrow} \quad \overset{\phi}{\Rrightarrow} \quad \overset{a'}{\searrow} \\ F(c) \xleftarrow{\;F(f)\;} F(c') \\ c \xrightarrow{\quad f \quad} c' \end{array} \right\}$$

Grothendieck-Riemann-Roch Theorem

$$K_{\overline{\mathcal{C}}_g/\overline{\mathcal{M}}_g} = c_1(\omega_{\overline{\mathcal{C}}_g/\overline{\mathcal{M}}_g})$$

$$\kappa_l = \pi_*(K^{l+1}_{\overline{\mathcal{C}}_g/\overline{\mathcal{M}}_g})$$

$$\mathbb{E} = \pi_*(\omega_{\overline{\mathcal{C}}_g/\overline{\mathcal{M}}_g})$$

$$\lambda_l = c_l(\mathbb{E})$$

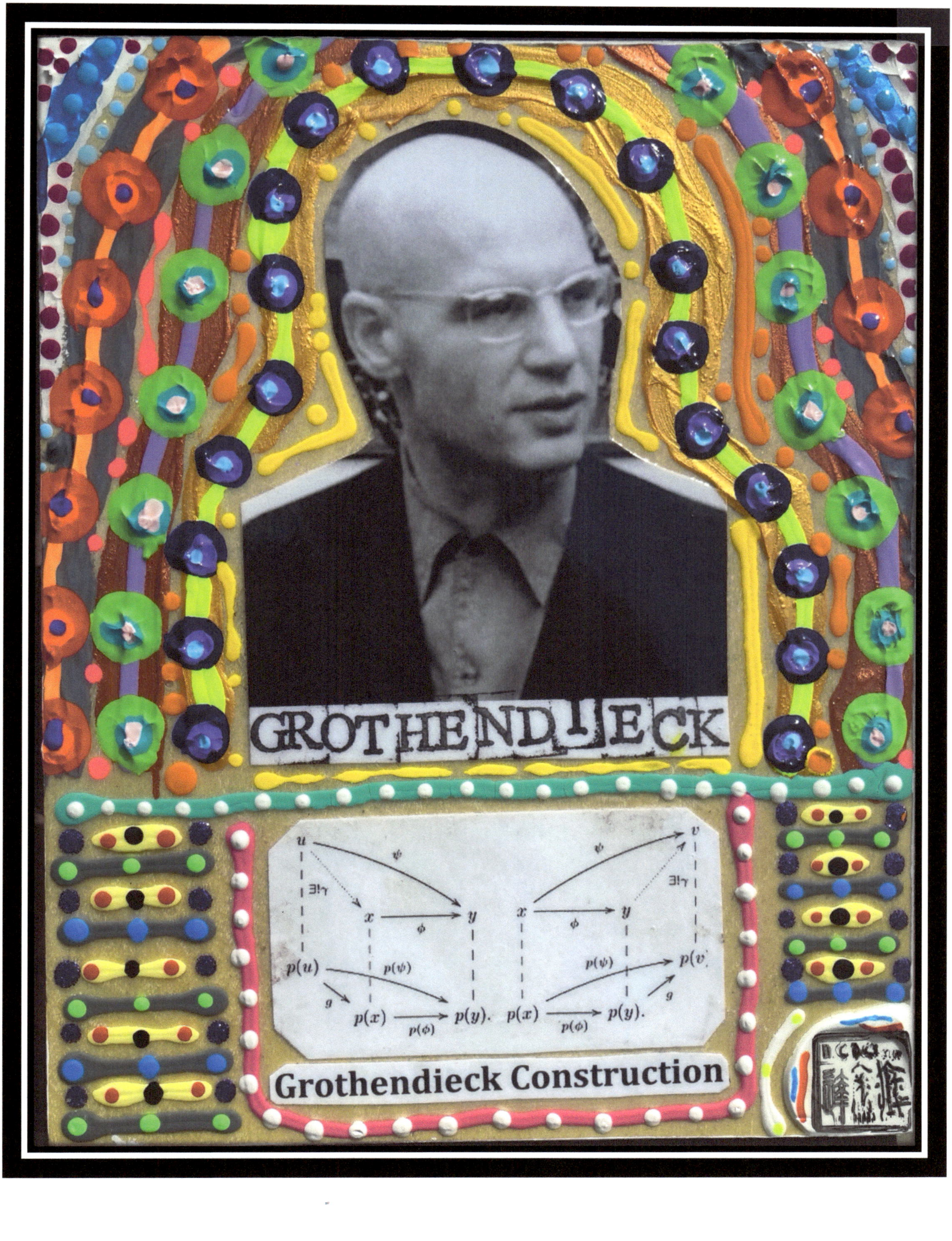
GROTHENDIECK
Grothendieck Construction

Murray Gell-Mann (1929-2019)
American of Jewish descent

In 1964, Gell-Mann theorized the existence of quarks, the subatomic particles that make up larger particles such as protons and neutrons, which are the constituents of atomic nuclei. Particles made up of quarks are called "hadrons". Thus the Large Hadron Collider (LHC) at CERN in Switzerland, which attempts to break hadrons apart. For some time, Gell-Mann was undecided on an actual spelling for the term he intended to coin. Then he found the word "quark" in Irish novelist James Joyce's 1939 book *Finnegans Wake*:

– Three quarks for Muster Mark!
Sure he hasn't got much of a bark
And sure any he has it's all beside the mark.

The American physicist George Zweig proposed the existence of the same model of hadrons as Gell-Mann at the exact same time. Zweig called quarks "aces". Quarks were first observed experimentally at the Stanford Linear Accelerator Center in California.
Gell-Mann was born in Manhattan to a modest Jewish immigrant family from what is now Ukraine. He excelled academically from an early age. He went on to have a supremely distinguished career. He was a professor at Cal Tech for almost 40 years. He made many, many contributions to the theory of elementary subatomic particles, and was awarded the Nobel Prize in Physics in 1969.

The Eightfold Way

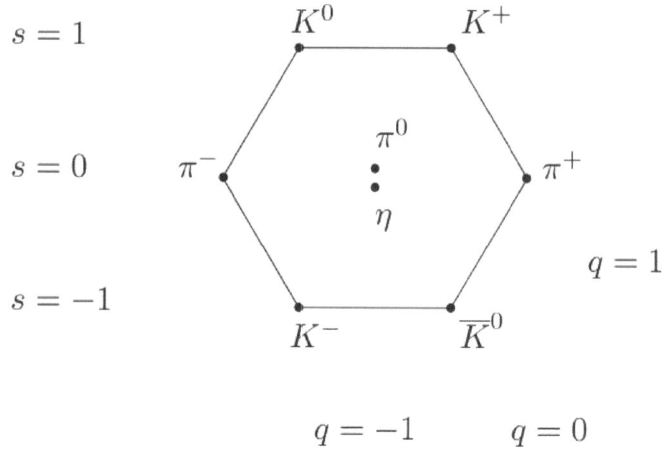

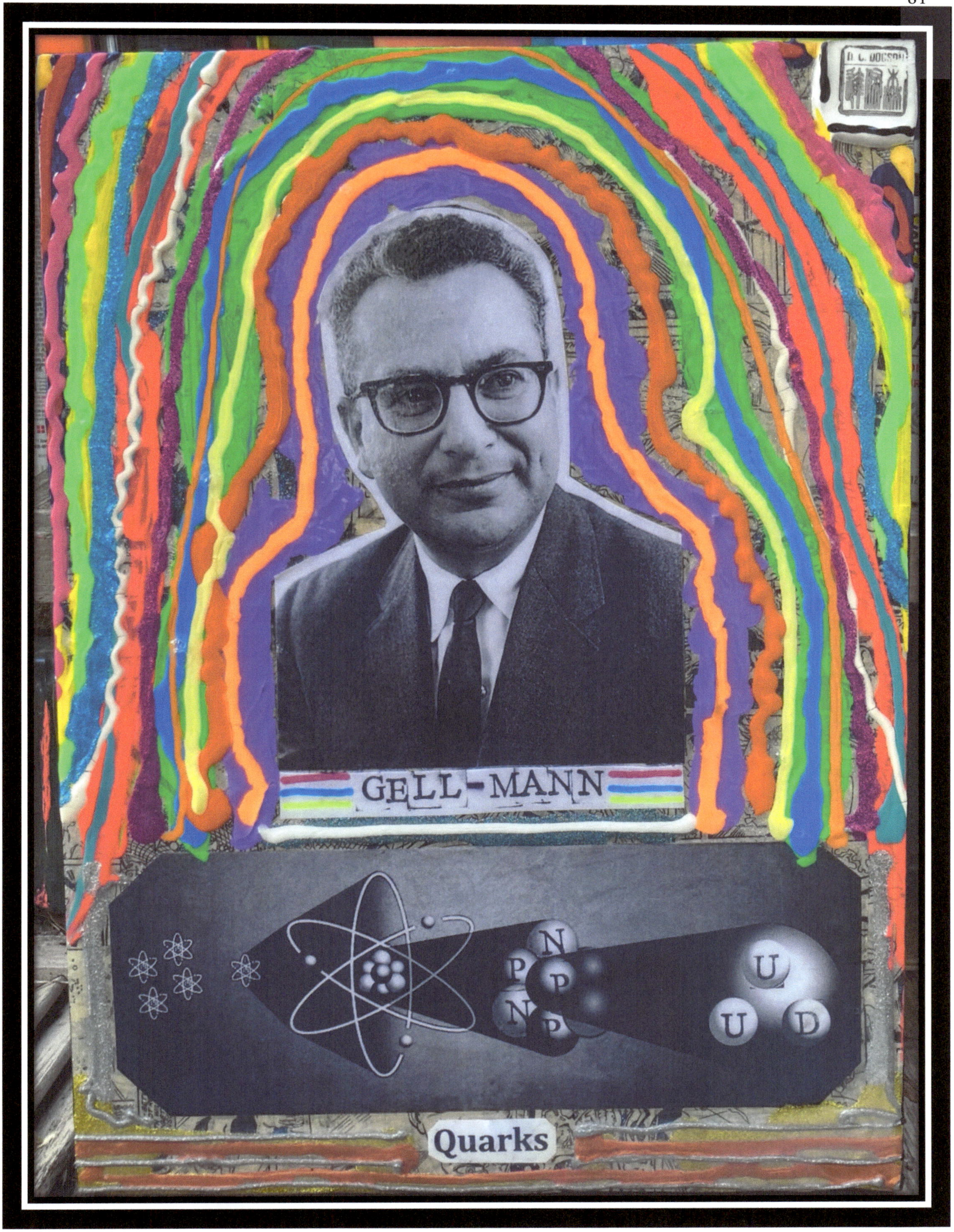
GELL-MANN
N
P
P
N
P
U
U
D
Quarks

Benoit Mandelbrot (1924-2010)

Polish-French American of Jewish descent

The Mandelbrot Set is the result of multiple recursive iterations of a simple equation on the complex plane. It is only obtainable using a computer. Fractals. Chaos Theory. The Mandelbrot Set represents an early fusion of mathematics and the computer. Today, of course, in 2024, computers are indispensable in mathematics, and rightly so. The striking thing about the Mandelbrot Set is its very organic appearance. In fact, many fractals resemble such organic phenomena as coastlines, trees, rivers and mountains. Mandelbrot was born in Warsaw, Poland, but moved to Paris when he was 11 years old. His parents were of Lithuanian Jewish extraction. His family somehow survived in Nazi-occupied France. Later, he became a dual French and American citizen. Mandelbrot spent much of his career at the IBM labs in New York, but also continued to work in France. Later in his life, he focused his attention on fields of applied mathematics such as economics, fluid dynamics and information theory.

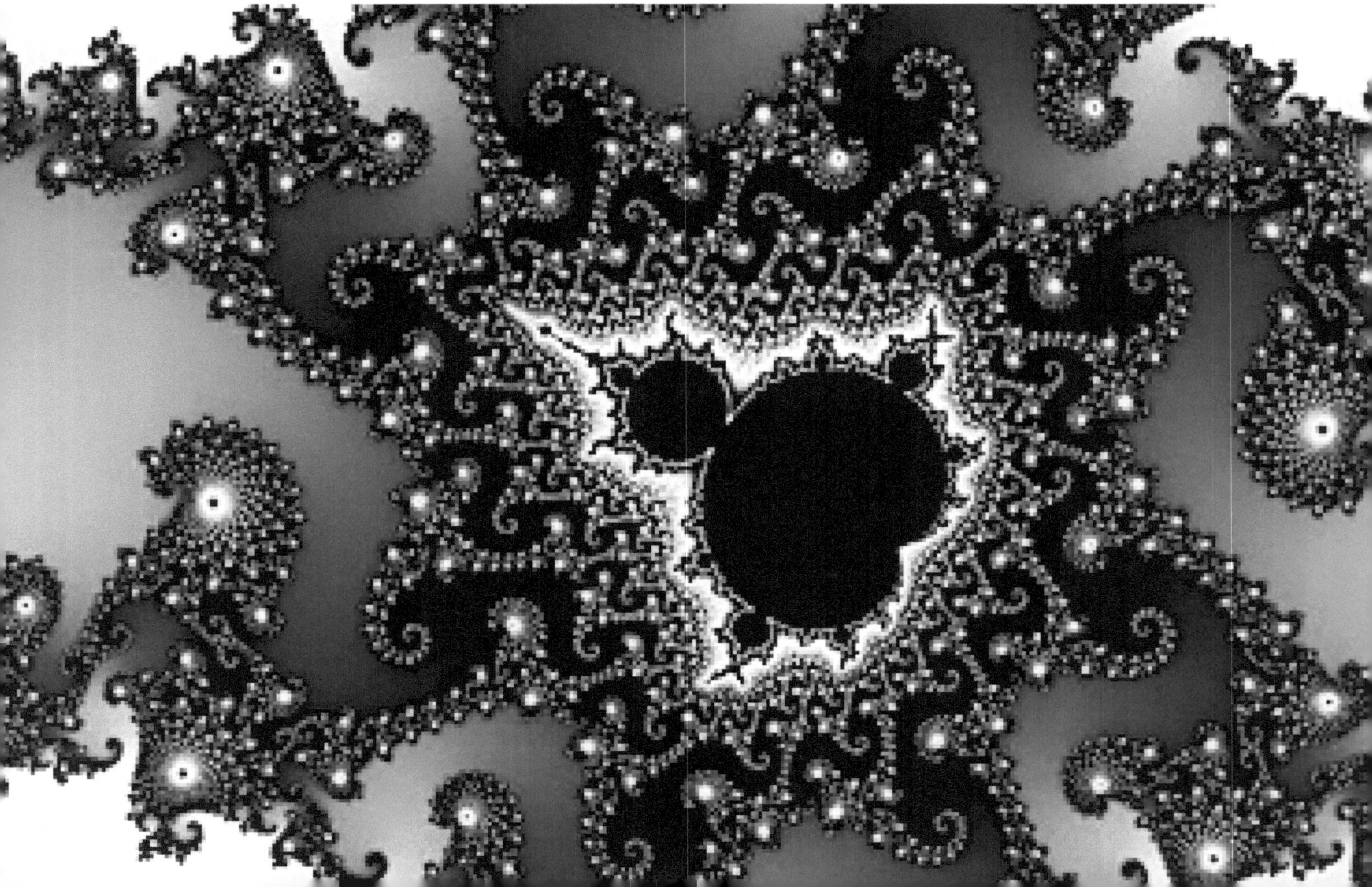

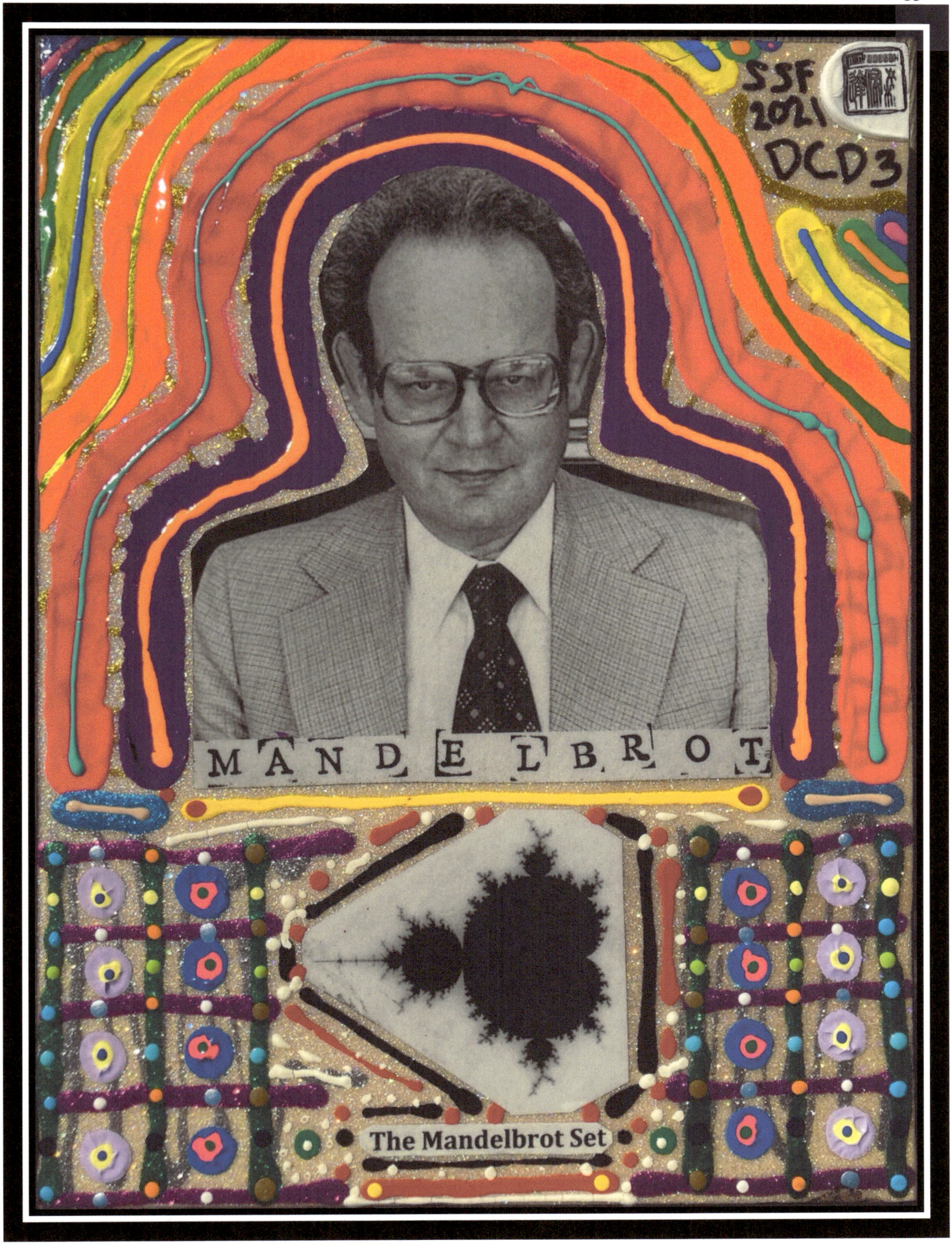
SSF
2021
DCD3
MANDELBROT
The Mandelbrot Set

The Electroweak Force: Sheldon Glashow (1932-), Steven Weinberg (1933-2021), Abdus Salam (1926-1996)

American Jewish (Glashow and Weinberg) and Pakistani (Salam)

In modern physics, there are said to be four fundamental forces of nature: gravity, the strong nuclear force, electromagnetism and the weak nuclear force. The ultimate goal of theoretical physics is to show that these forces can all be described by some overarching, unified scheme. After all, why should the very small operate under different laws from the very large?
These three showed that at high enough energies, such as existed just after the Big Bang, the electromagnetic force and the weak nuclear force act in unison. They become the same force: the electroweak force. For their work on this topic, these three were jointly awarded the Nobel Prize in Physics in 1979.

Sheldon Glashow was born in New York City to a Russian Jewish immigrant family. He received his PhD in physics from Harvard University in 1959. After stints at Stanford and Cal Berkeley, he became a professor of physics at Harvard in 1966. He became a professor emeritus at Harvard in 2000.

Steven Weinberg was also born into a Jewish immigrant family in New York City, at about the same time as Glashow. As it turns out, both he and Glashow went to the Bronx High School of Science and graduated the same year, 1950. They went their separate ways but reconnected later in life. Weinberg taught at Harvard, MIT and Columbia. In 1982, he relocated to the University of Texas at Austin, where he remained until his death.

Abdus Salam was born in present-day Pakistan, before the 1947 partition of British India. He received his PhD in theoretical physics from Cambridge University in England. He returned to Pakistan and worked as a scientific advisor to the government. He attempted to encourage the study of science and physics in Pakistan, and in all the so-called third world.

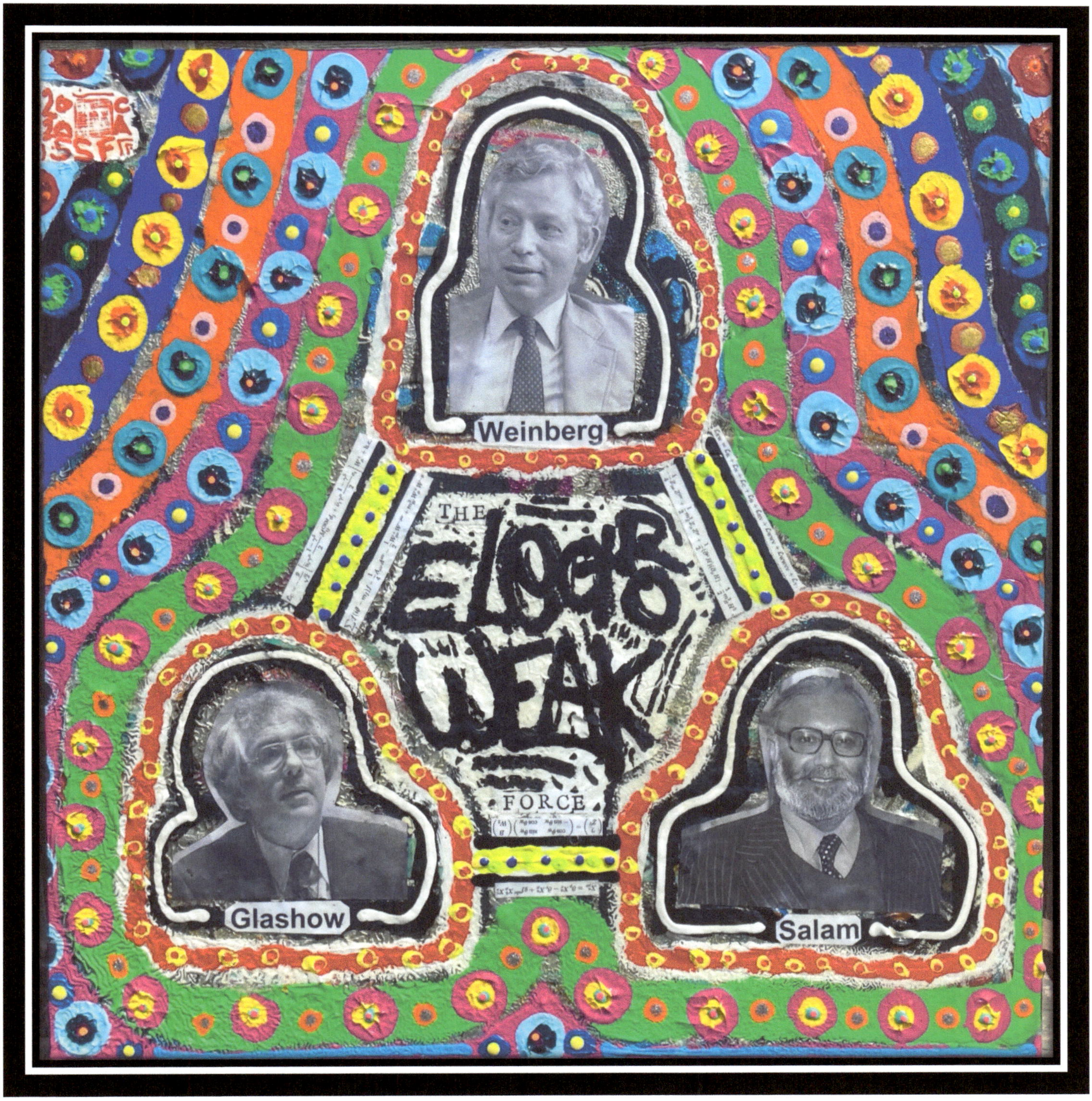
Weinberg
THE
FORCE
Glashow
Salam

John Conway (1937-2020)
English

Surreal Numbers are
Just numbers. Conway's
Game of Life is there.

The Monster Group and
Monstrous Moonshine are
Easy to dismiss.

John Conway was born and raised in Liverpool, England, and went on to study mathematics at Cambridge. In 1986, he moved to New Jersey, USA, to become the John von Neumann Professor of Mathematics at Princeton University. He held the post until his death. He passed away in 2020 at the age of 82 due to complications from Covid-19.

Conway's Game of Life

Go to playgameoflife.com and play it yourself.

1. A live cell dies if it has fewer than two live neighbors.
2. A live cell with two or three live neighbors lives on to the next generation.
3. A live cell with more than three live neighbors dies.
4. A dead cell is brought back to life if it has exactly three live neighbors.

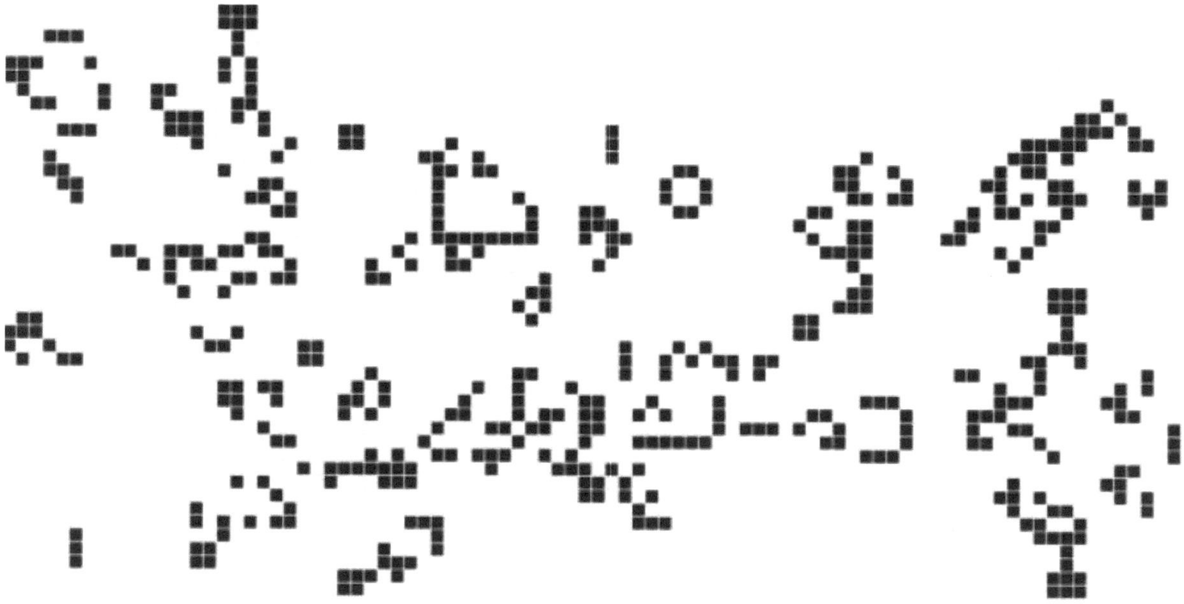

A screenshot from an iteration of Conway's Game of Life.

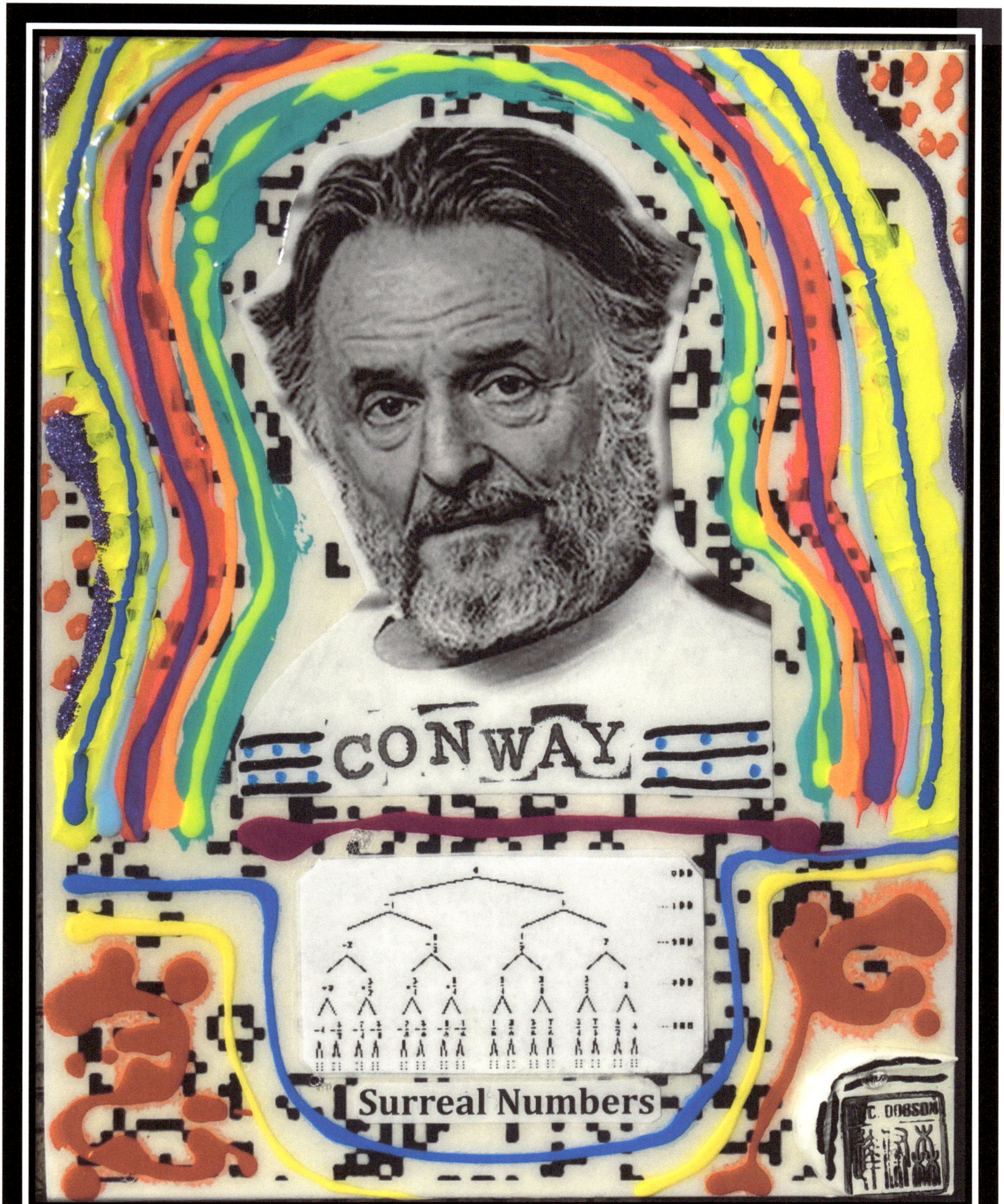
CONWAY
Surreal Numbers
C. DOBSON

Alain Connes (1947-)

French

Fields Medal, 1982

Born in a small town in southeastern France called Draguignan, Connes went on to study in Paris. He is a professor at the Collège de France and THE Ohio State University, among others.

Alain Connes is responsible for Noncommutative Geometry, which I always thought had a nice ring to it. It sounds to me like the world. If you don't know what commutativity is, you must have been asleep in 6th-grade math. The Commutative Property of algebra is that you can multiply numbers in any order and get the same result. Thus,

$$2 \times 4 = 4 \times 2$$

$$5 \times 10 \times 2 = 2 \times 5 \times 10$$

$$ab = ba$$

$$abc = bca = cab$$

That's commutativity, which is often grouped with the properties of associativity and distributivity. But commutativity is not a necessary property of algebra. There are algebras that are non-commutative, in which case order matters. Such is the case with time, which occurs in a single order and cannot be reversed. Geometry being the study of objects in space, Noncommutative Geometry sounds to me exactly like the world of objects in time.

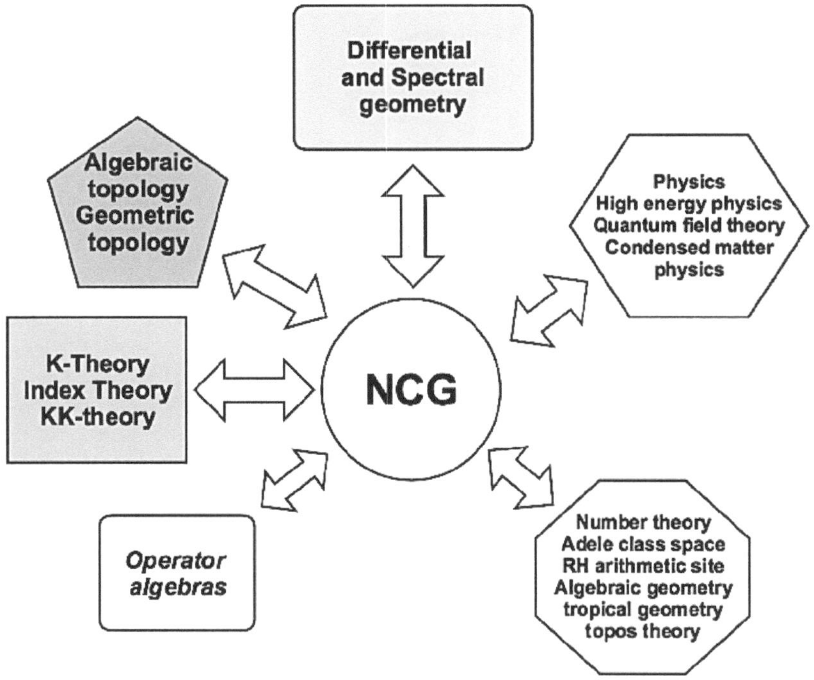

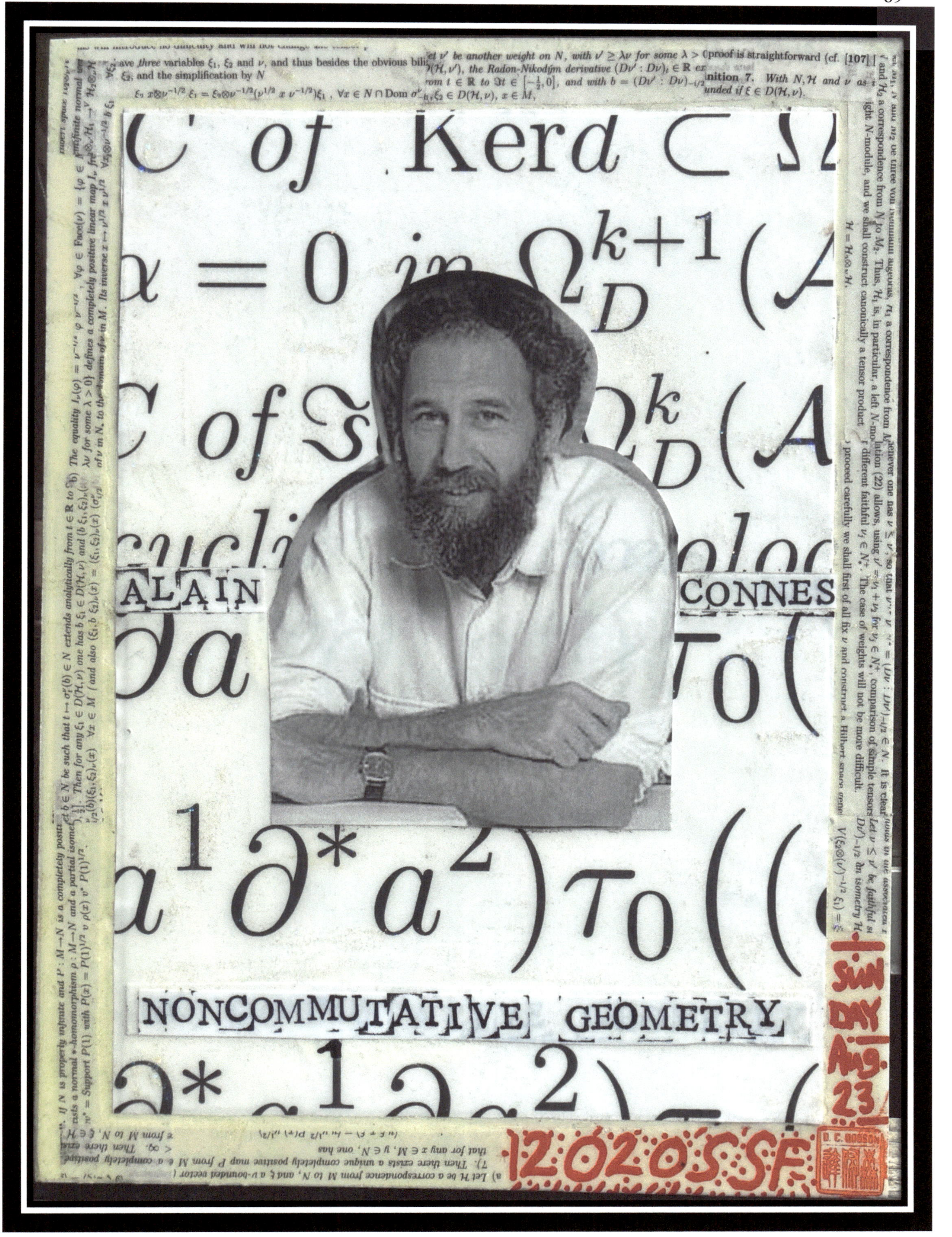
ALAIN
CONNES
NONCOMMUTATIVE GEOMETRY
SUN DAY Aug. 23
12 02 05 SF
D. C. HOBSON

Andrew Wiles (1953-)
English

In the 1990s, Andrew Wiles proved Fermat's Last Theorem, a notable outstanding problem in mathematics for 300 years or so. Professor Wiles teaches and does mathematical research at the University of Oxford in England.

If you have never heard of Fermat's Last Theorem, or don't understand it, then take note:

For right triangles, we know that $a^2 + b^2 = c^2$.
Are there integers a, b, and c for which $a^3 + b^3 = c^3$ or $a^4 + b^4 = c^4$?
Are there any integers *n* greater than 2 for which $a^n + b^n = c^n$?

In the 17th century, Fermat said no. More of a conjecture than a theorem, but Fermat said he could prove it, though there is no evidence of this.

In 1993, Wiles presented his initial proof to a small lecture hall at the University of Cambridge. The crowd was gobsmacked. Some errors were found in the proof, but Wiles corrected them and was totally vindicated by 1995.

In his proof, Wiles used mathematics that Fermat definitely did not have access to. In fact, Wiles exploited many findings that were state of the art at the time. The proof itself also provided many new issues for mathematicians to address.

The mathematical text here states Fermat's Last Theorem:
"There does not exist any x, y, z and n in the integers and zero such that n is greater than 2 and $x^n + y^n = z^n$."

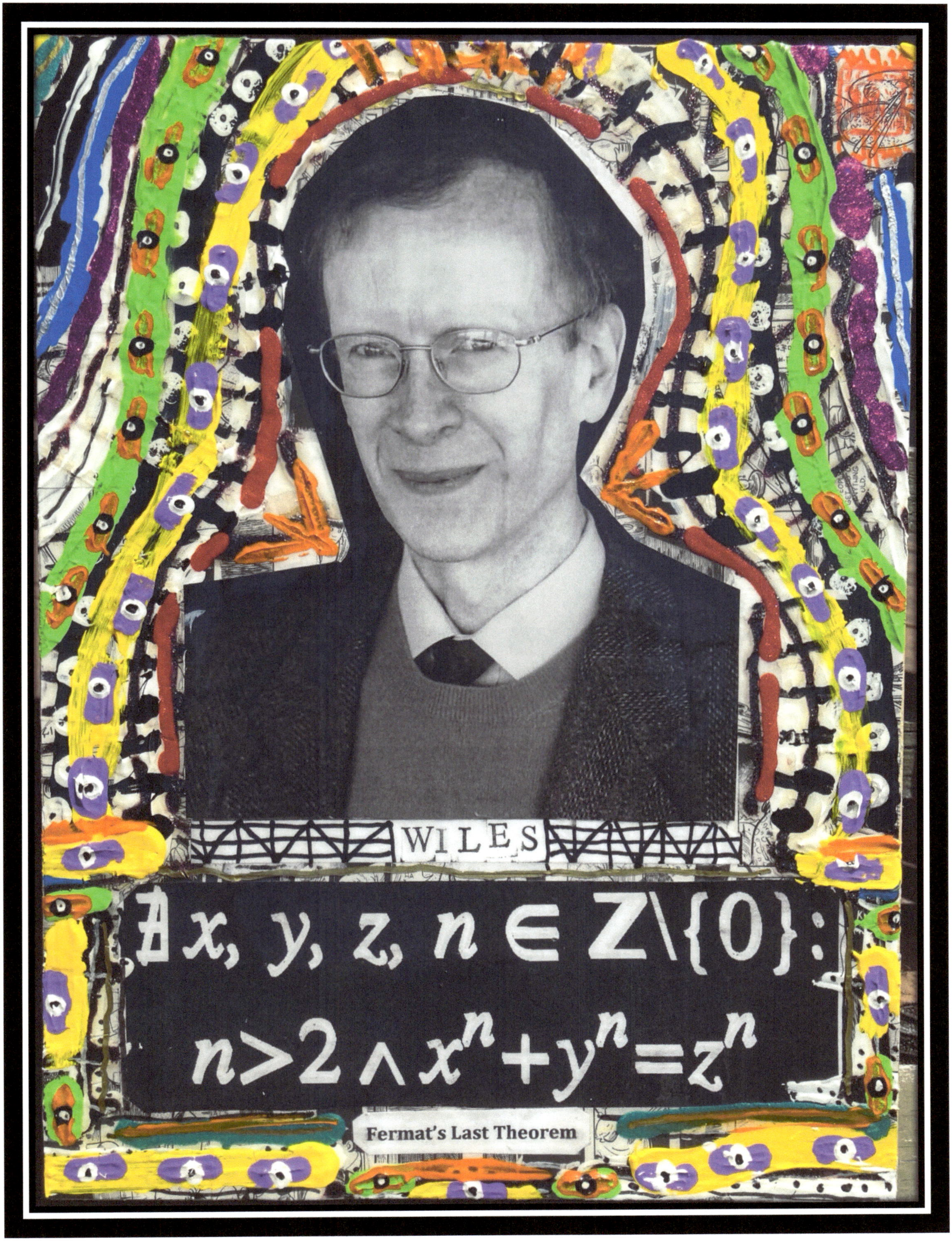
WILES
$\nexists x, y, z, n \in \mathbb{Z}\setminus\{0\}:$
$n>2 \wedge x^n+y^n=z^n$
Fermat's Last Theorem

Edward Witten (1951-) and Grigori Perelman (1966-)
American (Witten) and Russian (Perelman) of Jewish descent

In papers published in 2002 and 2003, Perelman proved the Poincaré Conjecture. In the 19th century, Poincaré conjectured that if it is possible to shrink any loop on a four-dimensional surface smoothly down to a point, then that surface is topologically equivalent to a hypersphere (a four-dimensional sphere). It so happened that the Poincaré Conjecture was one of the million-dollar Millennium Prize Problems. The Clay Math Institute was ready to give Perelman the prize, but he refused it. He was also awarded a Fields Medal in 2006 but rejected it as well. He seems to have felt that the contributions of those who worked on the problem before him were just as important as his own, and he didn't want to be singled out. Still, it seems to me he could have provided some good attention and publicity for mathematics.

Witten, born in Baltimore, Maryland, was awarded a Fields Medal in 1990, primarily for the mathematics he developed to describe string theory. It was unusual to award the medal to a physicist. String theory is the idea that so-called fundamental particles actually extend into higher dimensions and, instead of being point-like, are in fact string-like in nature. The extra dimensions would be rolled or folded up so that we cannot perceive them. In theory, the strings could vibrate in certain different ways that would represent different particles. The problem with string theory is that it has been around since the 1980s and it hasn't proved a thing. 11 dimensions? 26 dimensions? No proof? Nothing? Sorry. String theory is a near-total failure, except for some interesting mathematics. As Wolfgang Pauli once said in response to a confused theory, "That is not only not right; it is not even wrong."

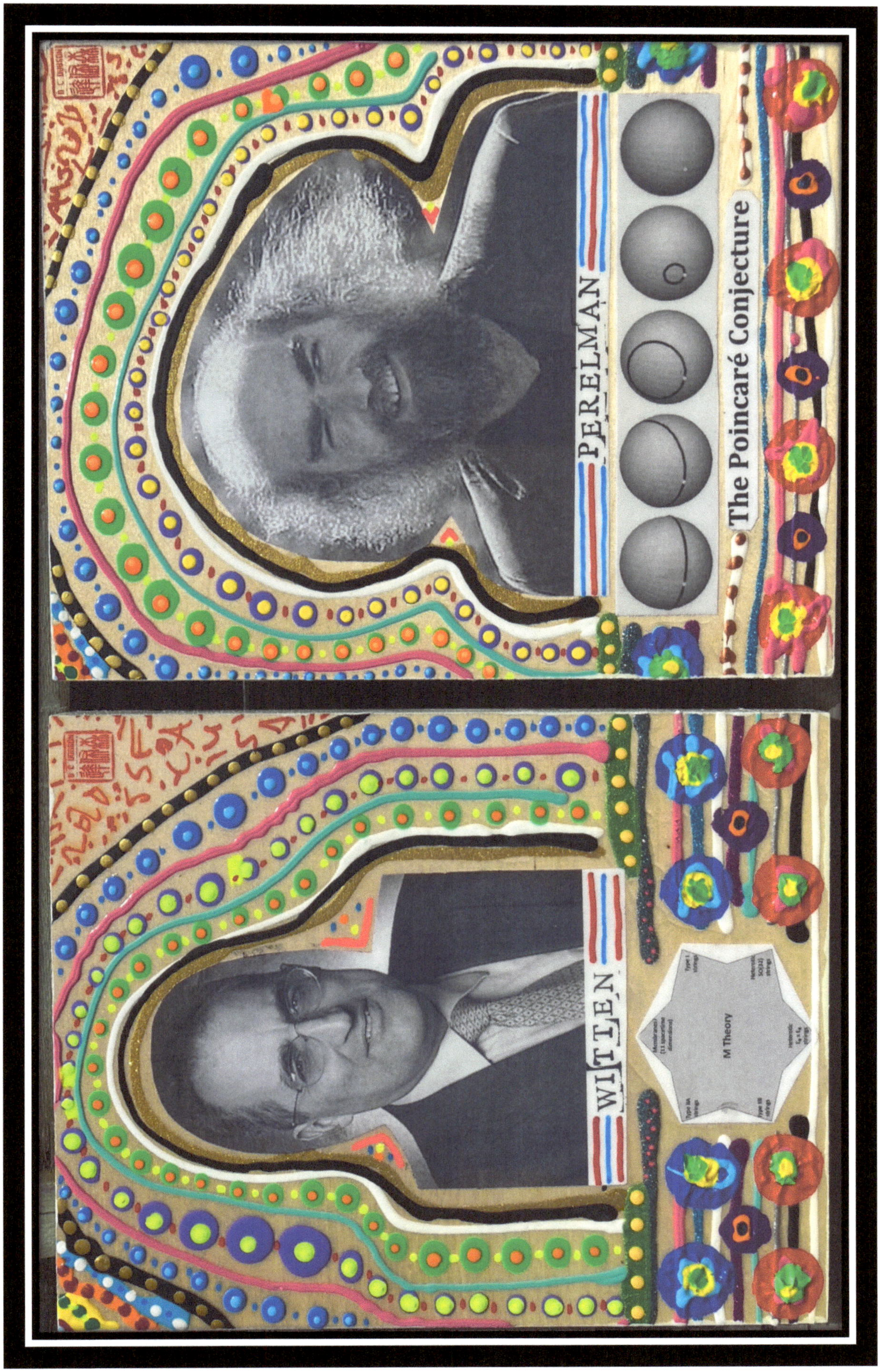
PERELMAN
The Poincaré Conjecture
WITTEN
M Theory

Sean M. Carroll (1966-)

American of mixed European descent

Sean Carroll is a contemporary philosopher, popularizer, and serious physicist and cosmologist. He was born in Philadelphia, PA. He received his PhD in astronomy from Harvard University in 1993. He is well known for his personal atheistic philosophy, which he has called "poetic naturalism." He has called science "methodological empiricism," as opposed to methodological naturalism.

The World of Everyday Experience

In One Equation

quantum mechanics — spacetime — gravity

$$W = \int_{k<\Lambda} [Dg][DA][D\psi][D\Phi] \exp\left\{ i \int d^4x \sqrt{-g} \left[\frac{m_p^2}{2} R - \frac{1}{4} F^a_{\mu\nu} F^{a\mu\nu} + i\bar{\psi}^i \gamma^\mu D_\mu \psi^i + \left(\bar{\psi}^i_L V_{ij} \Phi \psi^j_R + \text{h.c.} \right) - |D_\mu \Phi|^2 - V(\Phi) \right] \right\}$$

other forces — matter — Higgs

"God is not described in equations."

-S. Carroll

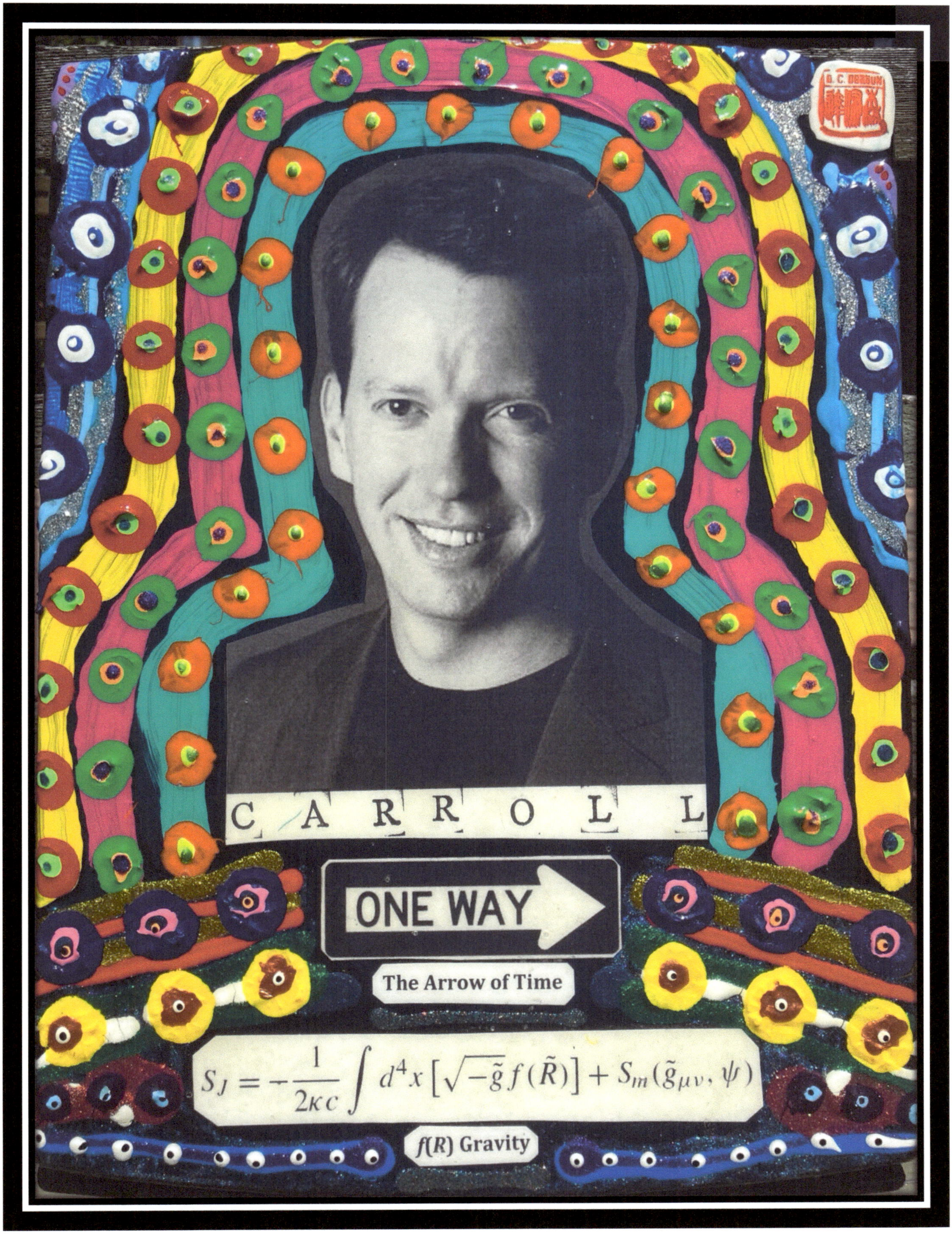
CARROLL
ONE WAY
The Arrow of Time
$S_J = -\frac{1}{2\kappa c}\int d^4x\left[\sqrt{-\tilde{g}}\,f(\tilde{R})\right] + S_m(\tilde{g}_{\mu\nu}, \psi)$
$f(R)$ Gravity

Terrence Tao (1975-)

Australian-American of Chinese descent

Terry Tao is a professor at UCLA. He received a Fields Medal in 2004, primarily for his work in number theory, including his proof that the prime numbers contain arithmetic progressions of any length. If you are wondering what that means, I will explain. What are prime numbers? The simple explanation is correct: a prime number is only evenly divisible by itself and 1. We are only considering integers here, whole counting numbers. Number Theory is mainly the study of integers, and in this study prime numbers play a key role. So what is the sequence of prime numbers as we count up the number line? Without going into too much detail, let us say that neither 0 nor 1 is prime. Thus, we have:

2, 3, 5, 7, 11, 13, 17, 19, 23, 29, 31, 37, 41, 43, 47, 53, 59, 61, 67, 71, 73, 79, 83, 89, 97, 101, 103, 107, 109, 113, ...

As an aside, note that it can be easily proven that there are infinitely many prime numbers. But what is the pattern of their progression? Is there a pattern? No one knows. Given the sequence up to the largest known prime number, can we say what the next one is? Nope. The fact is that after 2500 years or more of trying to answer these questions, no one knows the answers.

Now, the next largest prime number is found from time to time anyway through sheer brute-force computing involving many powerful computers all working together for years at a time. We are talking huge numbers that, were they expressed in our decimal system, would have millions upon millions of digits. The current largest known prime number, for example, is $2^{82,589,933} - 1$.

But what is desired in mathematics is a proof of the general case. In the case of prime numbers, there is no proof one way or the other. So is there anything we can say about the distribution of the prime numbers? Well, Terry Tao proved that it contains arithmetic progressions of any length. An arithmetic progression is a sequence of numbers separated by adding a fixed amount. There are innumerable known examples of such sequences among the primes, but Terry Tao proved the general case. Considering how little we know about the progression of the primes, this is a remarkable result.

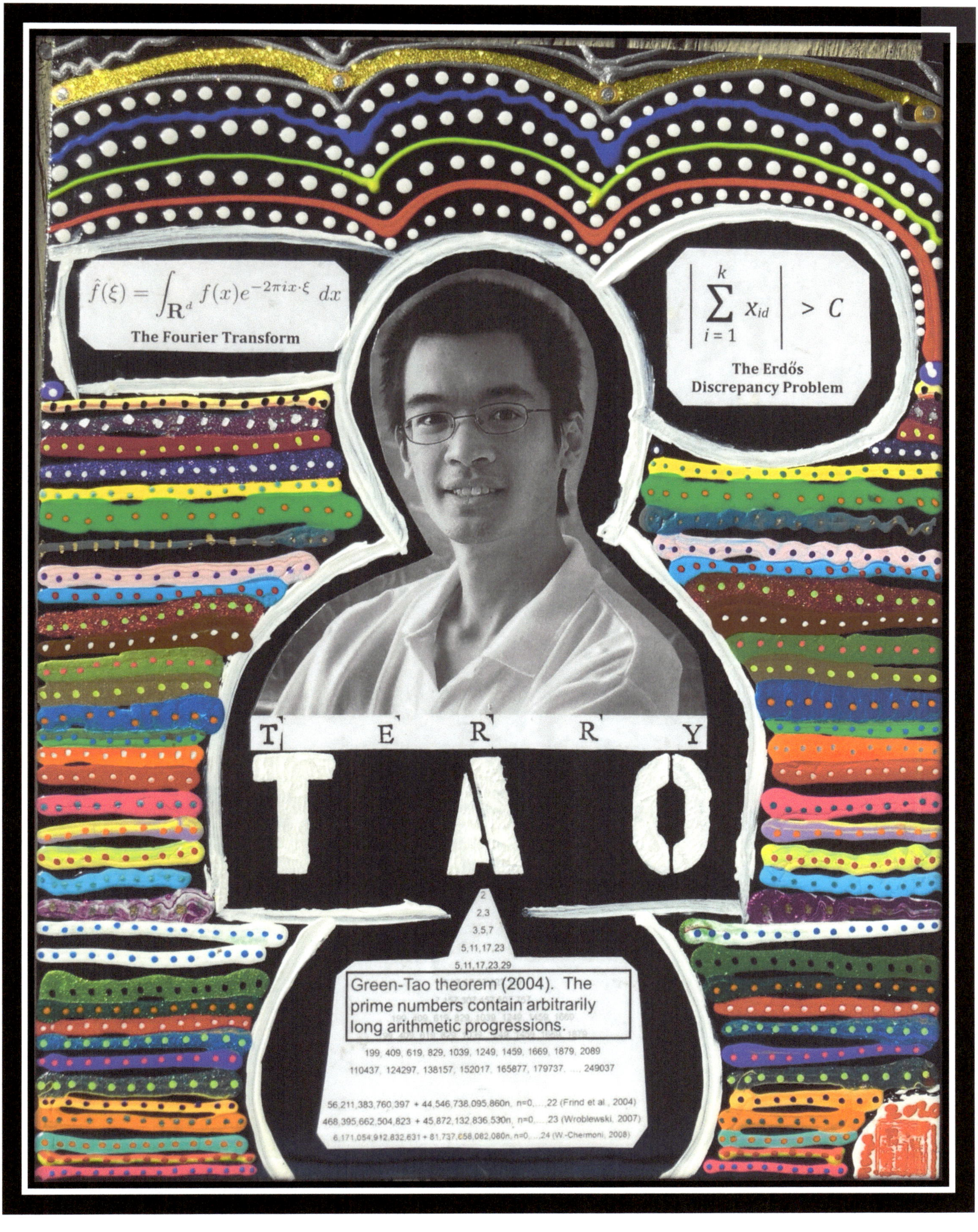
$\hat{f}(\xi) = \int_{\mathbf{R}^d} f(x) e^{-2\pi i x \cdot \xi} \, dx$
The Fourier Transform
$\left| \sum_{i=1}^{k} x_{id} \right| > C$
The Erdős Discrepancy Problem
TERRY
TAO
2
2,3
3,5,7
5,11,17,23
5,11,17,23,29
Green-Tao theorem (2004). The prime numbers contain arbitrarily long arithmetic progressions.
199, 409, 619, 829, 1039, 1249, 1459, 1669, 1879, 2089
110437, 124297, 138157, 152017, 165877, 179737, ..., 249037
56,211,383,760,397 + 44,546,738,095,860n, n=0,...,22 (Frind et al., 2004)
468,395,662,504,823 + 45,872,132,836,530n, n=0,...,23 (Wroblewski, 2007)
6,171,054,912,832,631 + 81,737,658,082,080n, n=0,...,24 (W.-Chermoni, 2008)

Maryam Mirzakhani (1977-2017)
Iranian

Fields Medal, 2014
Maryam Mirzakhani was a child prodigy in mathematics and was the first woman to be awarded a Fields Medal, as well as the first Iranian. She sadly passed away in 2017 at the age of 40 from breast cancer at Stanford Hospital in California. In this painting, we see some of her sketches.

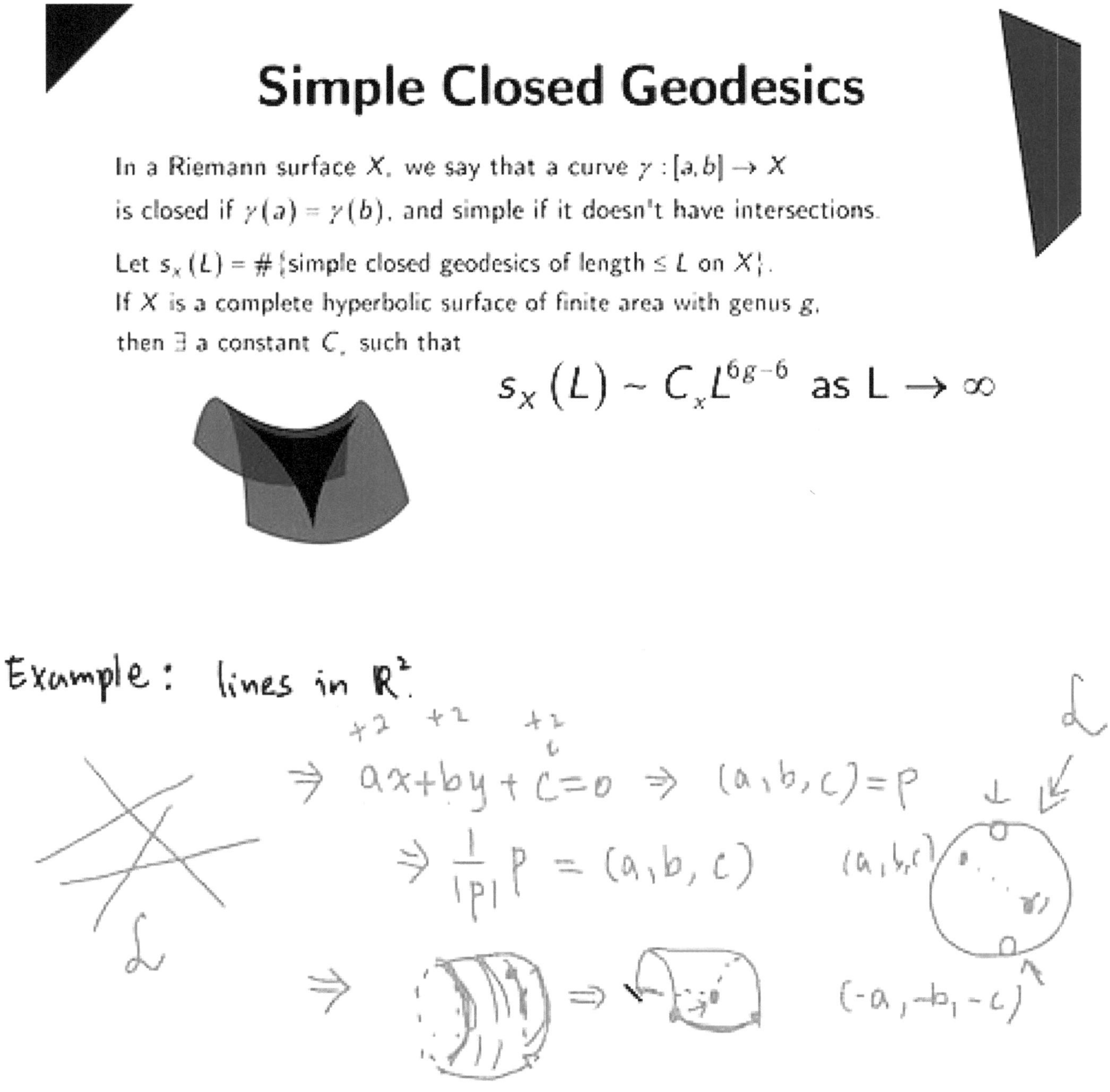

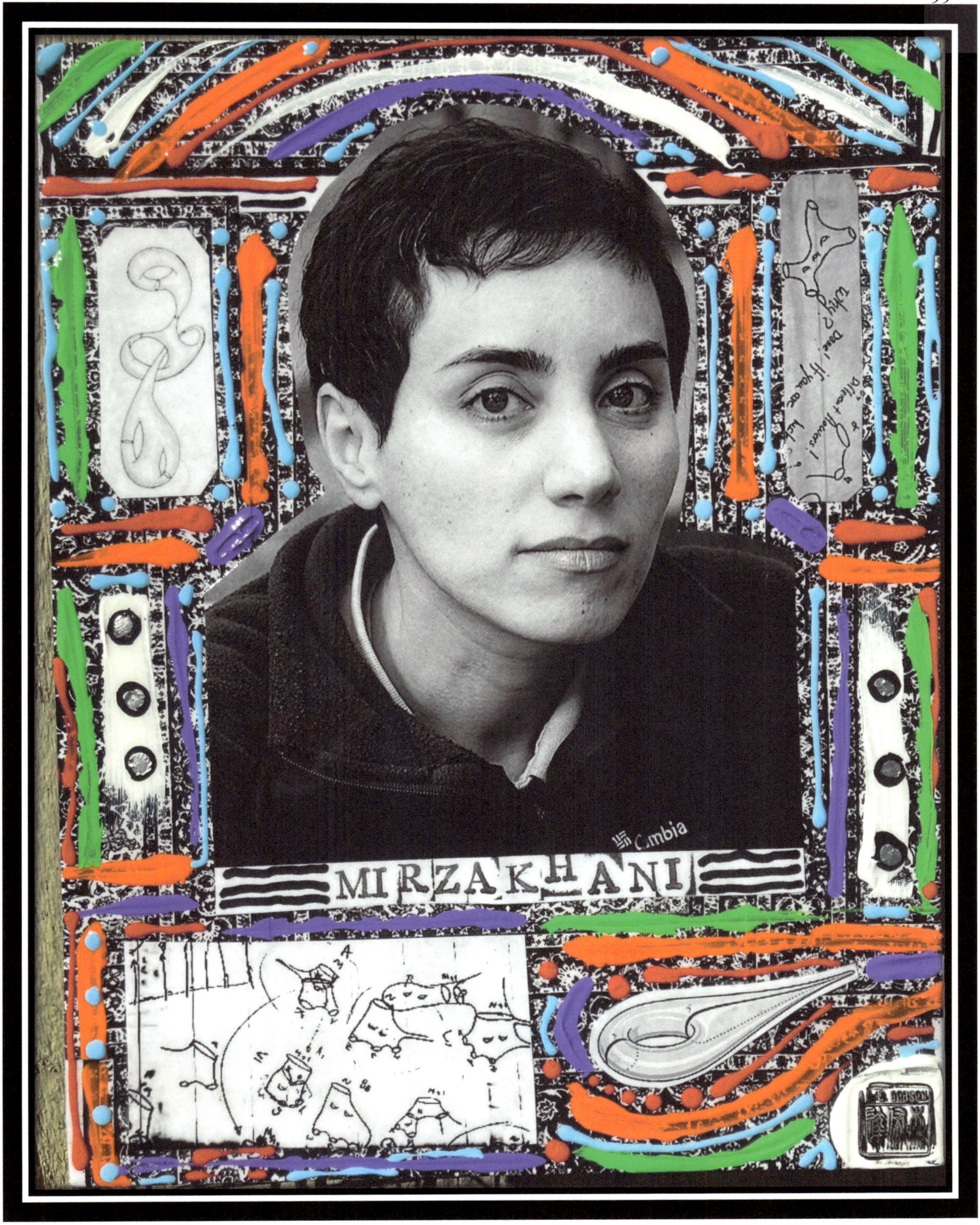
MIRZAKHANI

Manjul Bhargava (1974-)
Canadian American of Indian descent

Fields Medal, 2014
Manjul Bhargava was born in Ontario, Canada, but mostly went to school in New York. He is a math professor at Princeton University in New Jersey. His mother, Mira, was a math professor at Hofstra in NY, and taught him much about mathematics from an early age.

Bhargava Factorials

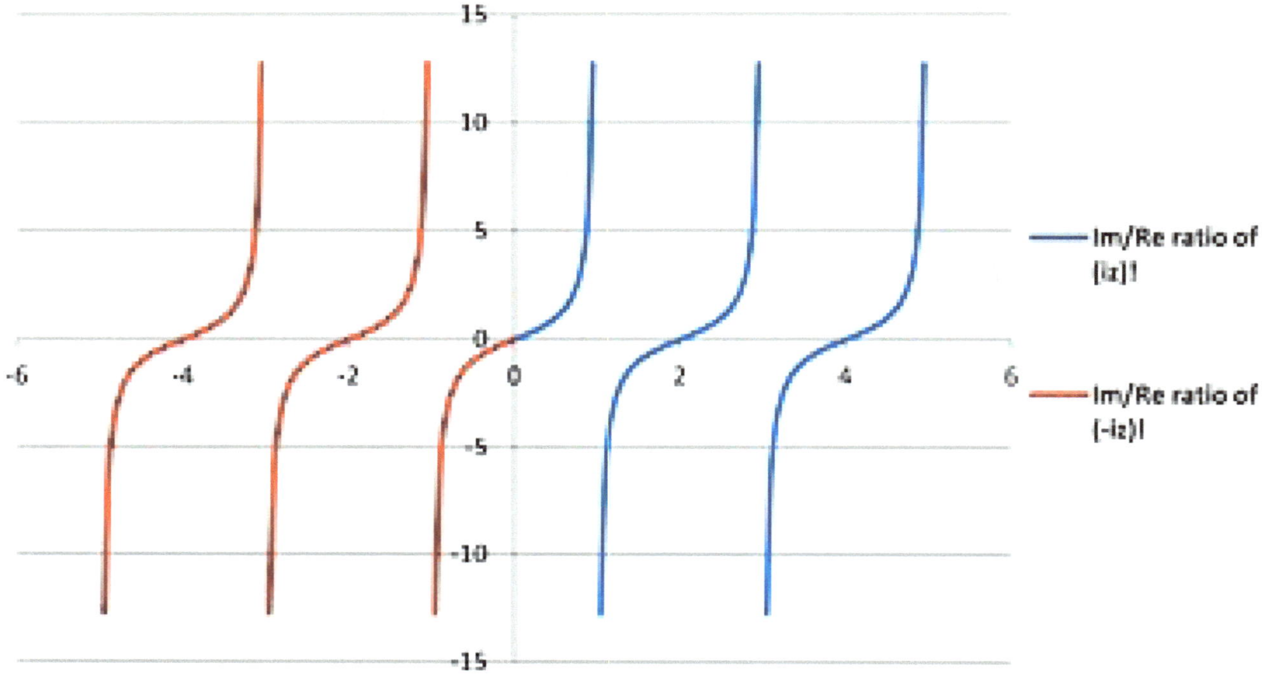

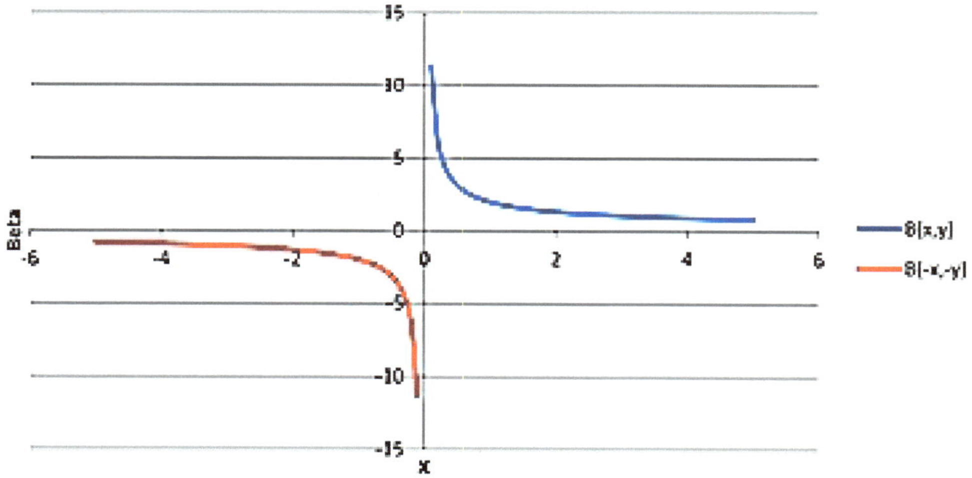

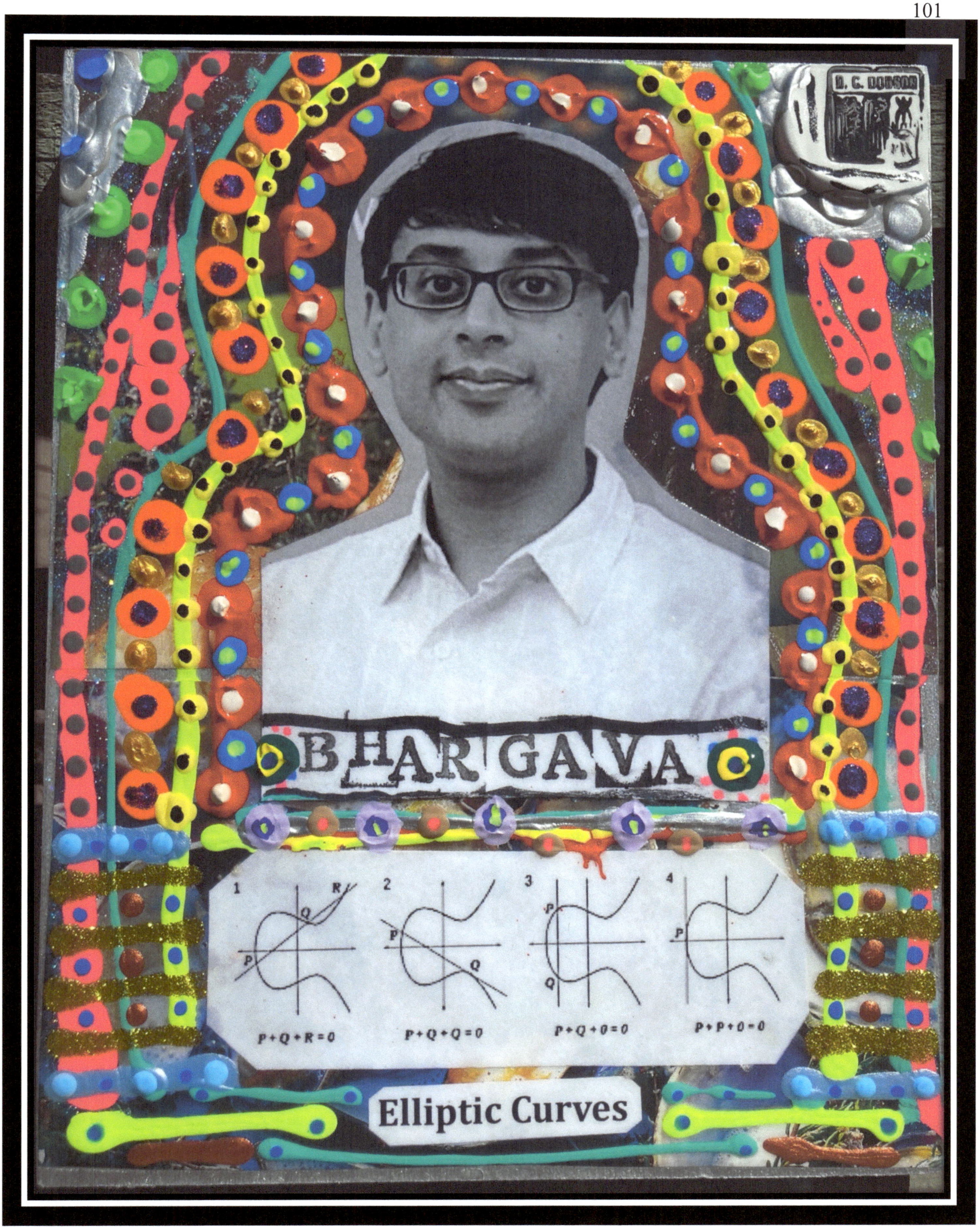
D. C. DOBSON
BHARGAVA
1
2
3
4
P+Q+R=0
P+Q+Q=0
P+Q+0=0
P+P+0=0
Elliptic Curves

James Maynard (1987-)
English

Fields Medal, 2022
Maynard was recognized for his results in Number Theory, such as showing that the prime numbers and the gaps between them must obey certain rules. For example, he proved that there are an infinite number of consecutive primes with a gap of 600. The goal is to bring this number down. Maynard's work greatly improved on that of his many predecessors. One cute result of his is the proof that for all decimal digits (0-9), there are infinitely many prime numbers that do not contain that digit. But, as I have noted, there are infinitely many prime numbers to begin with. Maynard proved this in 2016.

Maynard was born in Essex, England, and studied at both Cambridge and Oxford. He has received numerous other awards for his mathematical work. He was made a fellow of the UK's Royal Society in 2023. He is a professor at Oxford University in England.

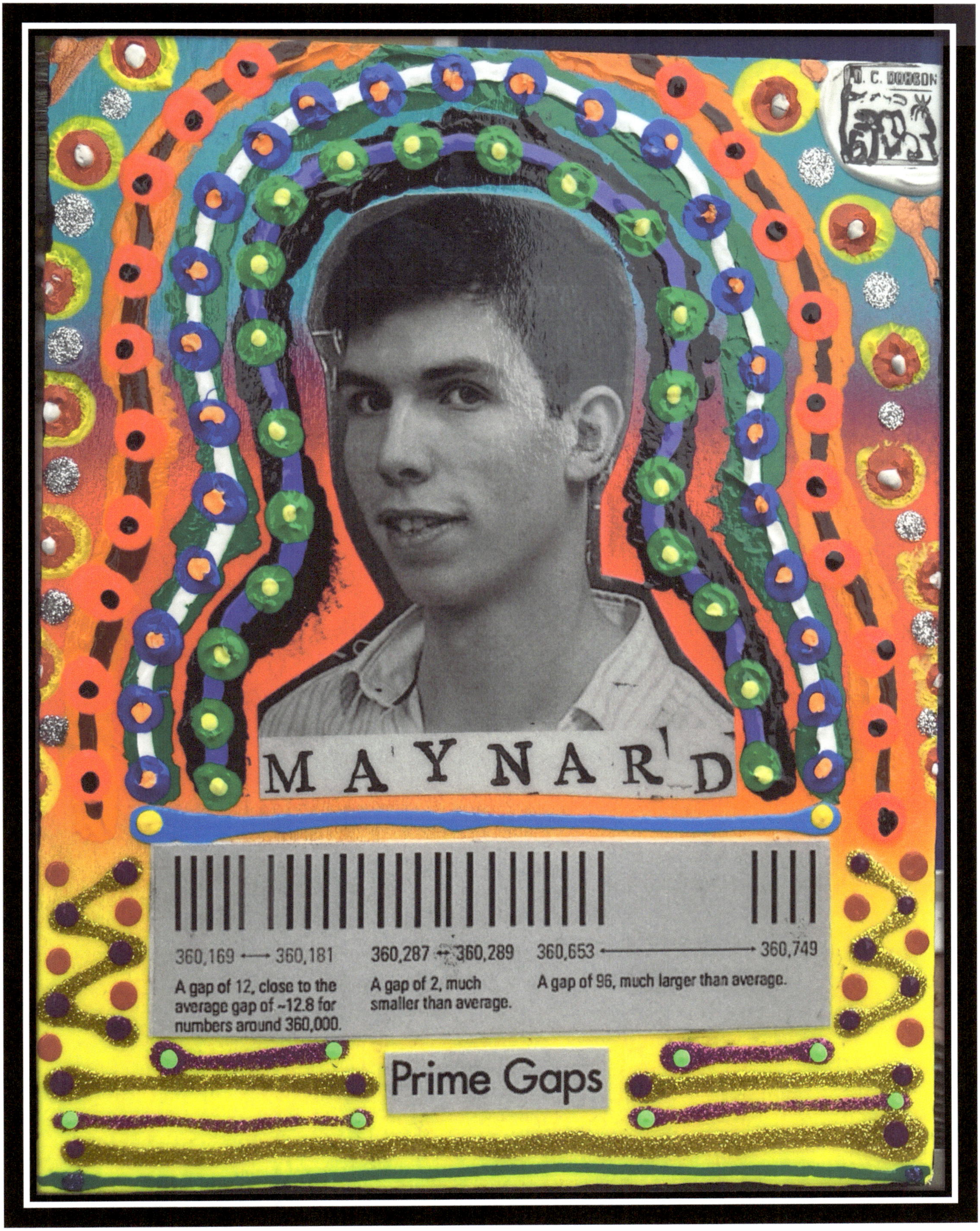
D. C. DOBSON
MAYNARD
360,169 ⟶ 360,181
A gap of 12, close to the average gap of ~12.8 for numbers around 360,000.
360,287 ↔ 360,289
A gap of 2, much smaller than average.
360,653 ⟶ 360,749
A gap of 96, much larger than average.
Prime Gaps

FIN

(for now...)

Appendices

These guys are not necessarily less important, but I didn't think of them in time. The project really could go on forever. As George Lucas said, "A movie is never finished, only abandoned." There is much more the reader can research for him or herself.

Appendix A

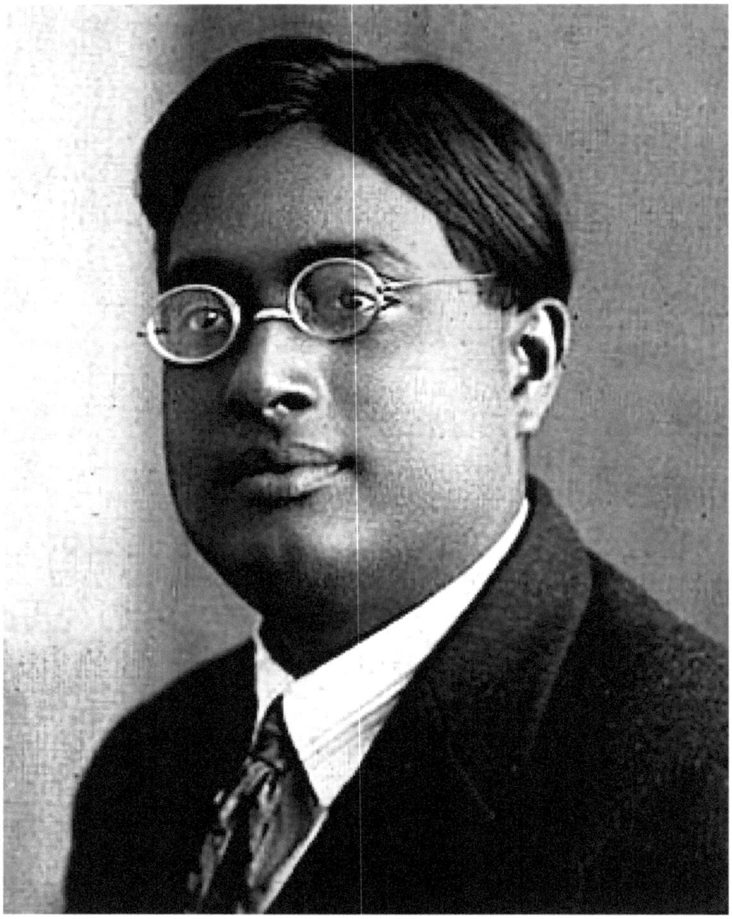

BOSE

(1894 – 1974)

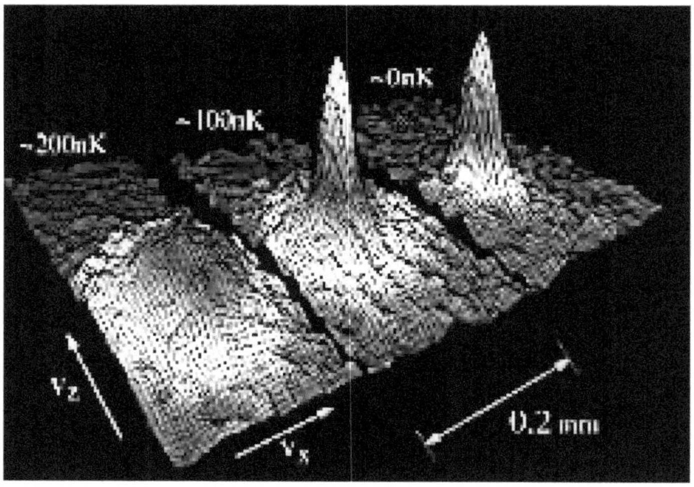

Bose-Einstein Condensate

Appendix B

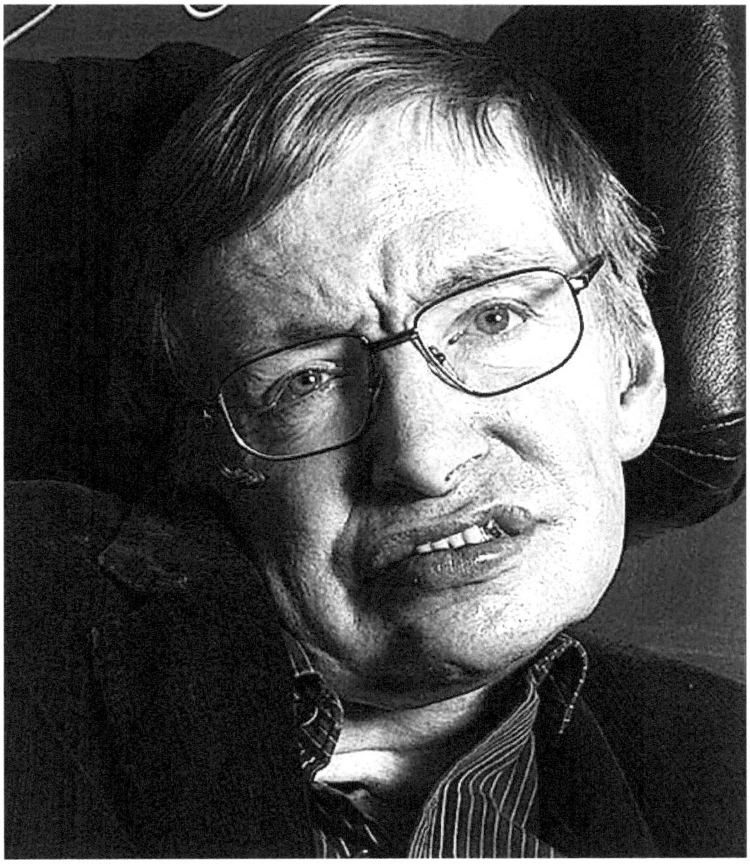

HAWKING

(1942 – 2018)

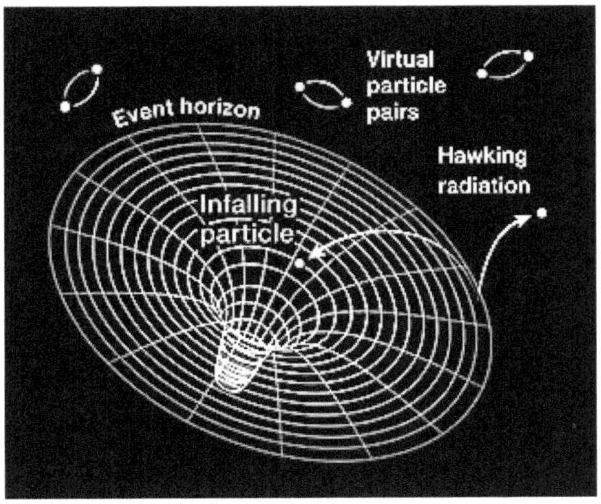

Hawking Radiation

Appendix C

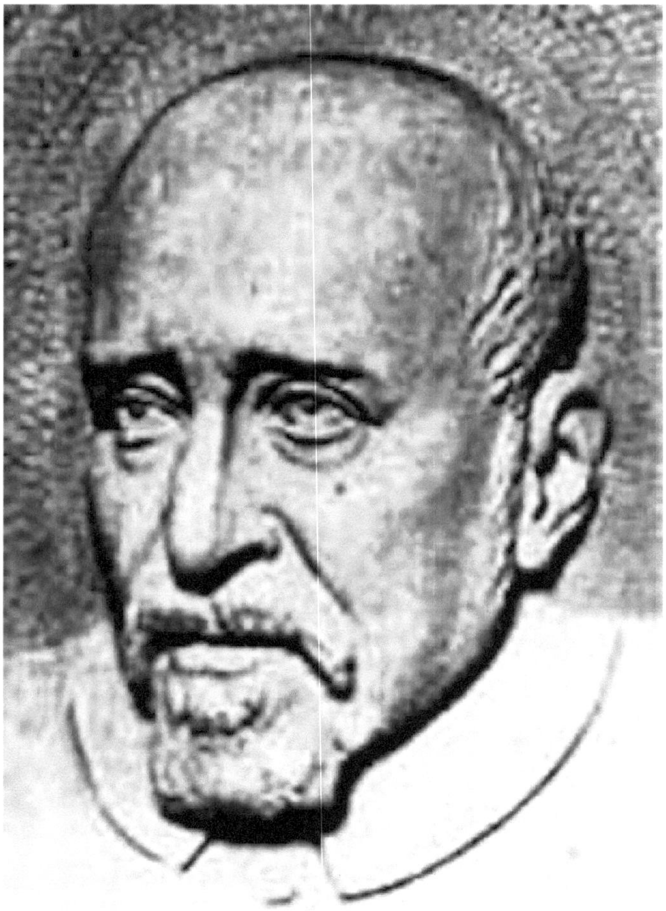

RECORDE

(c. 1510 – 1558)

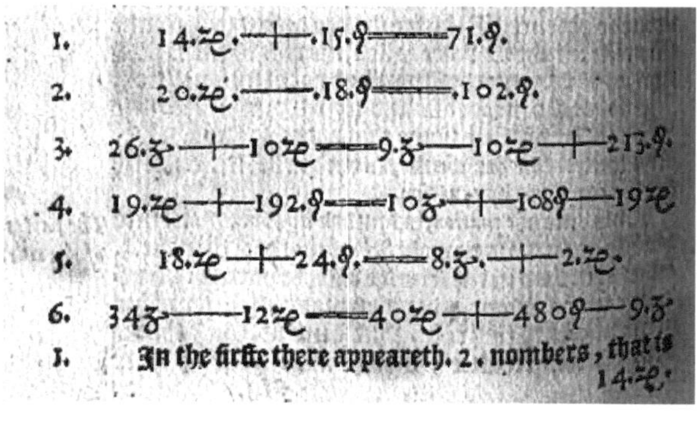

Plus and Equals Signs

Appendix D

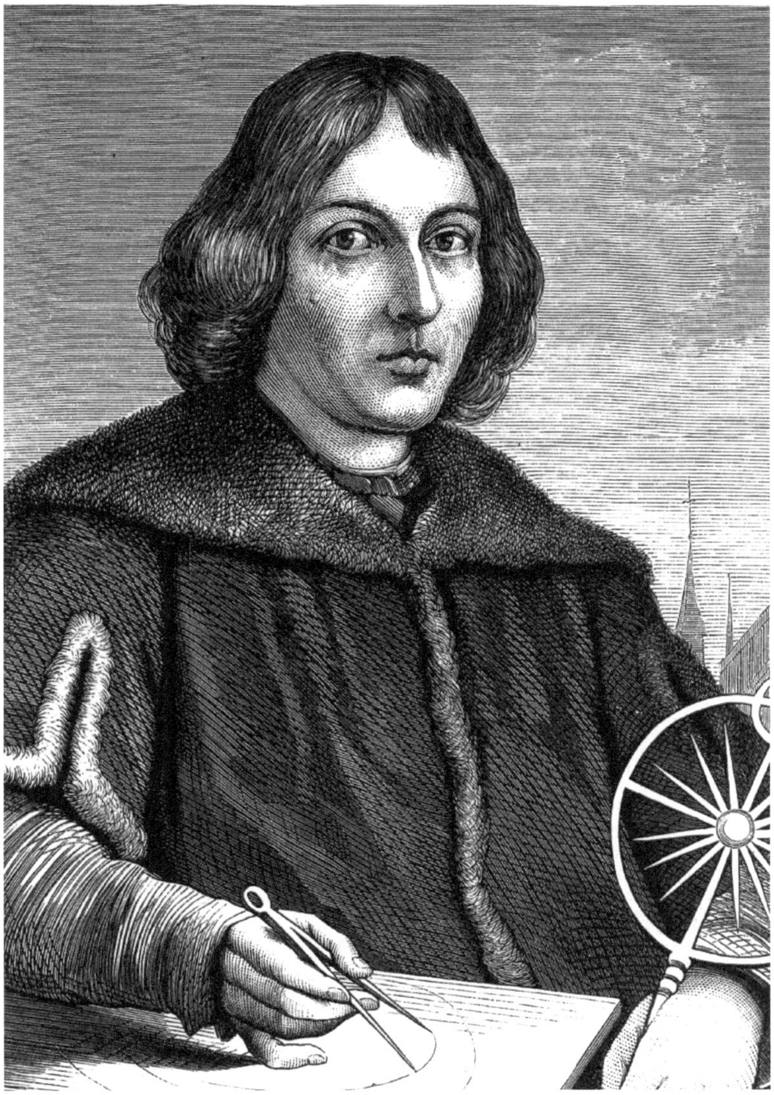

COPERNICUS

(1473 – 1543)

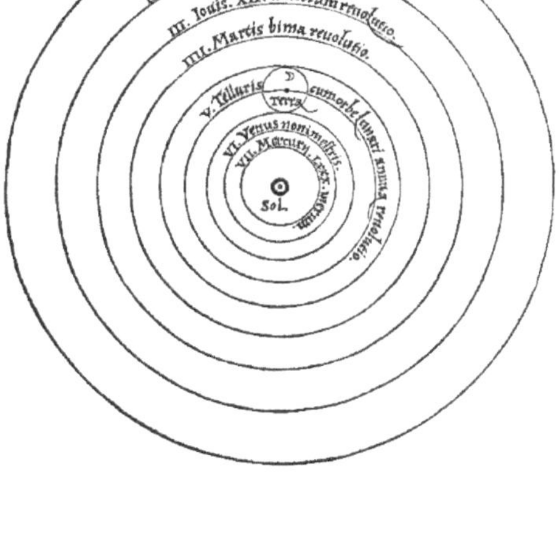

Appendix E

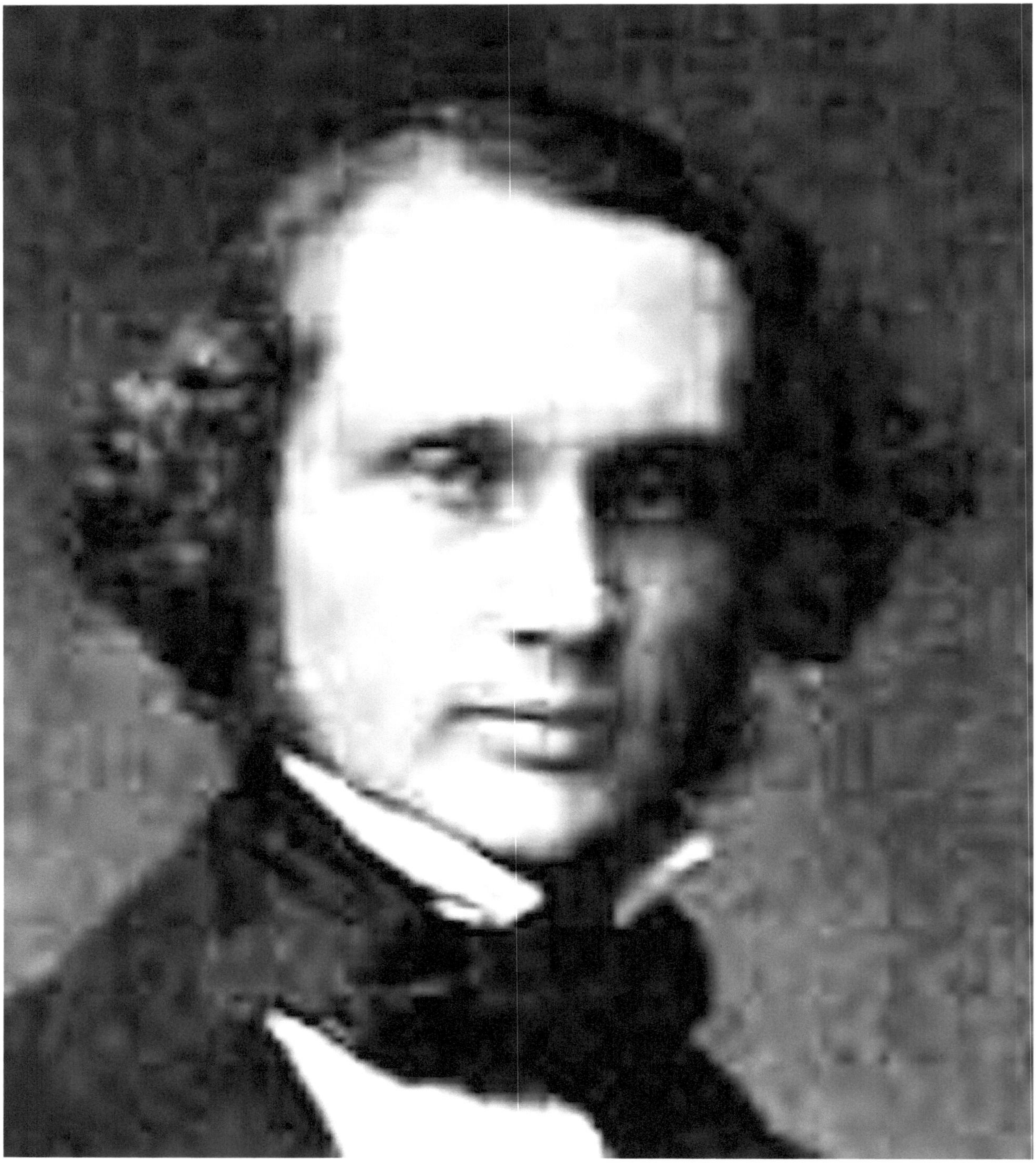

Lord Kelvin

(1824 – 1907)

Appendix F

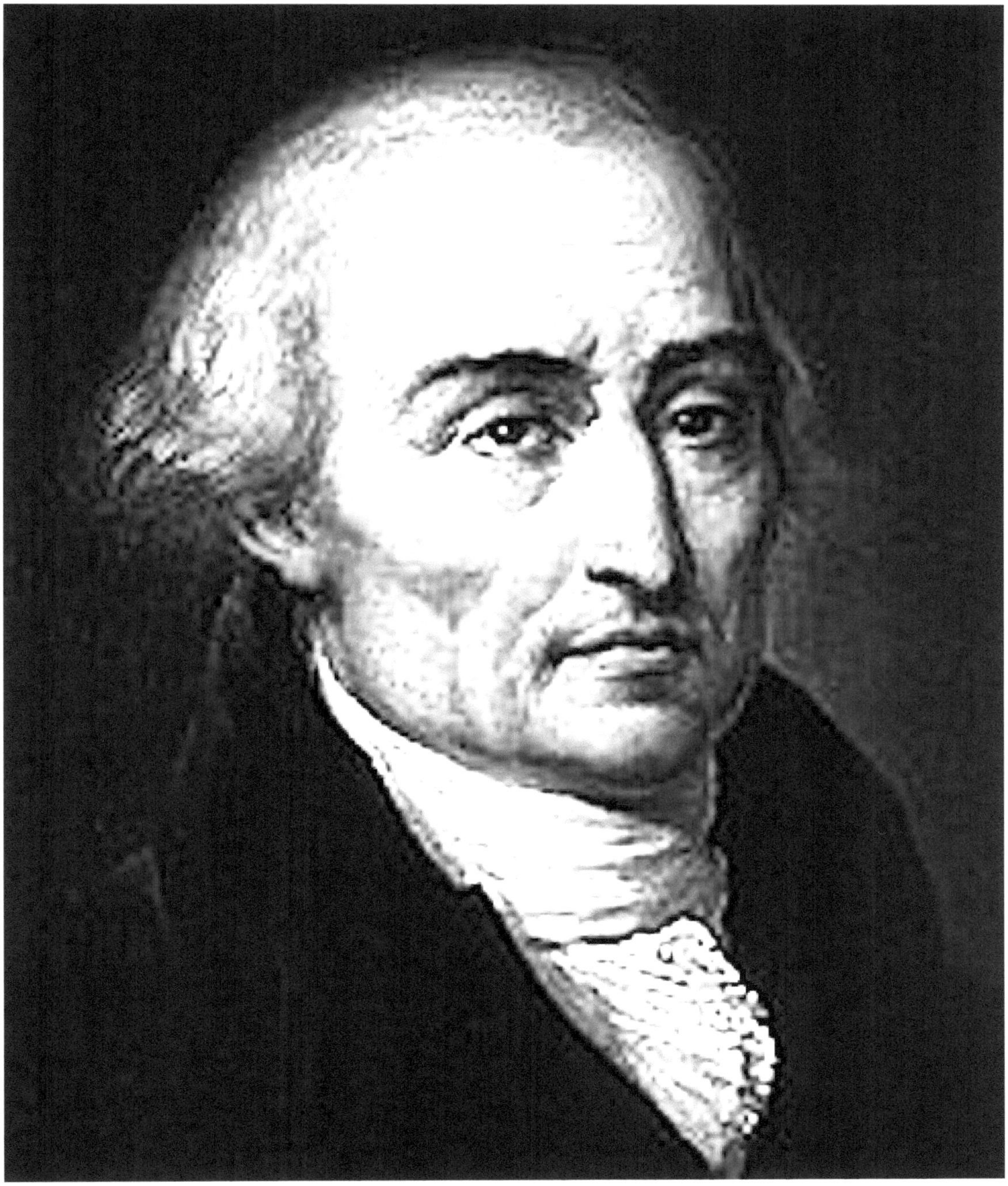

Lagrange

(1736 – 1813)

Appendix G

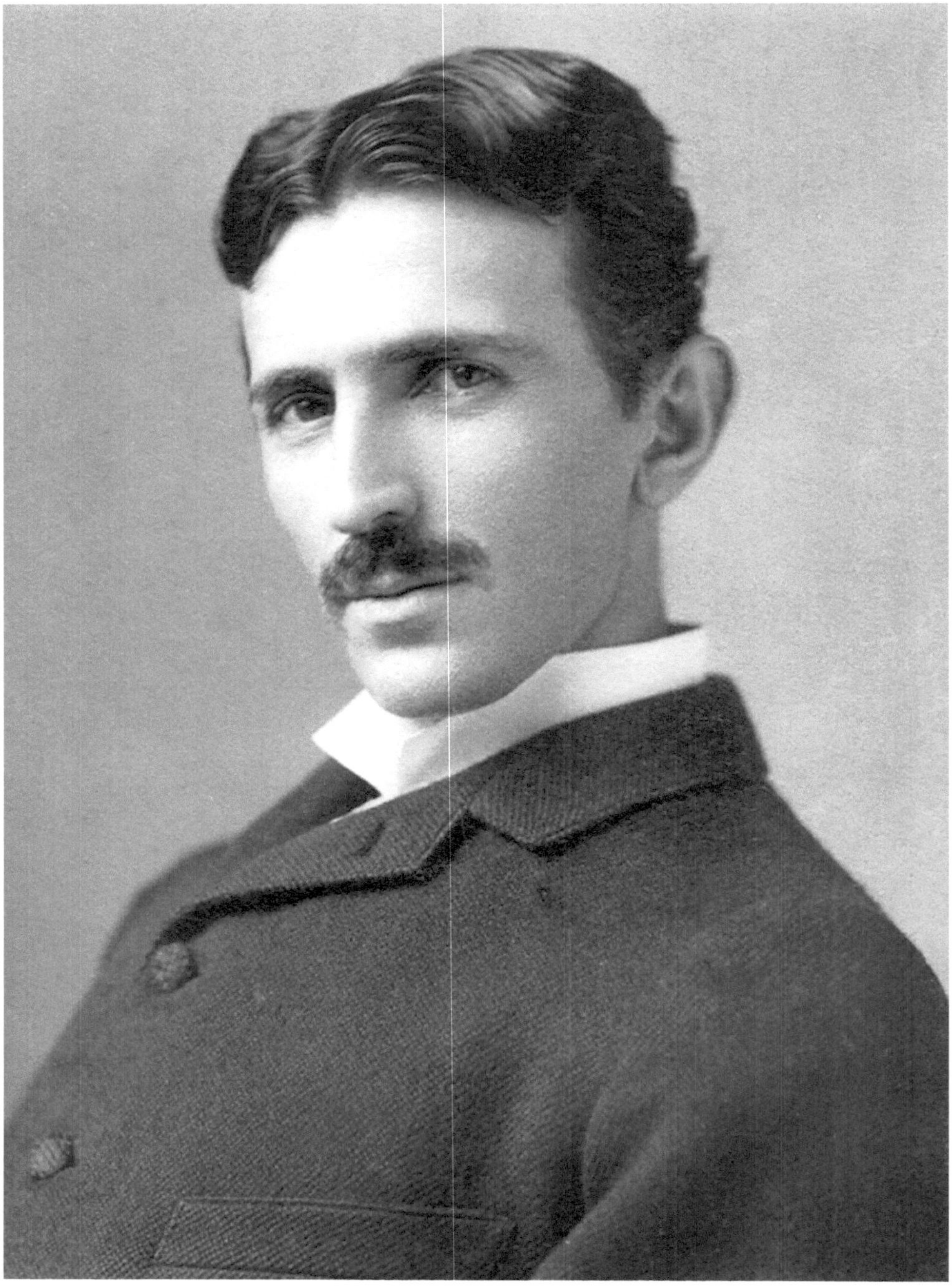

TESLA

(1856 – 1943)

Appendix H

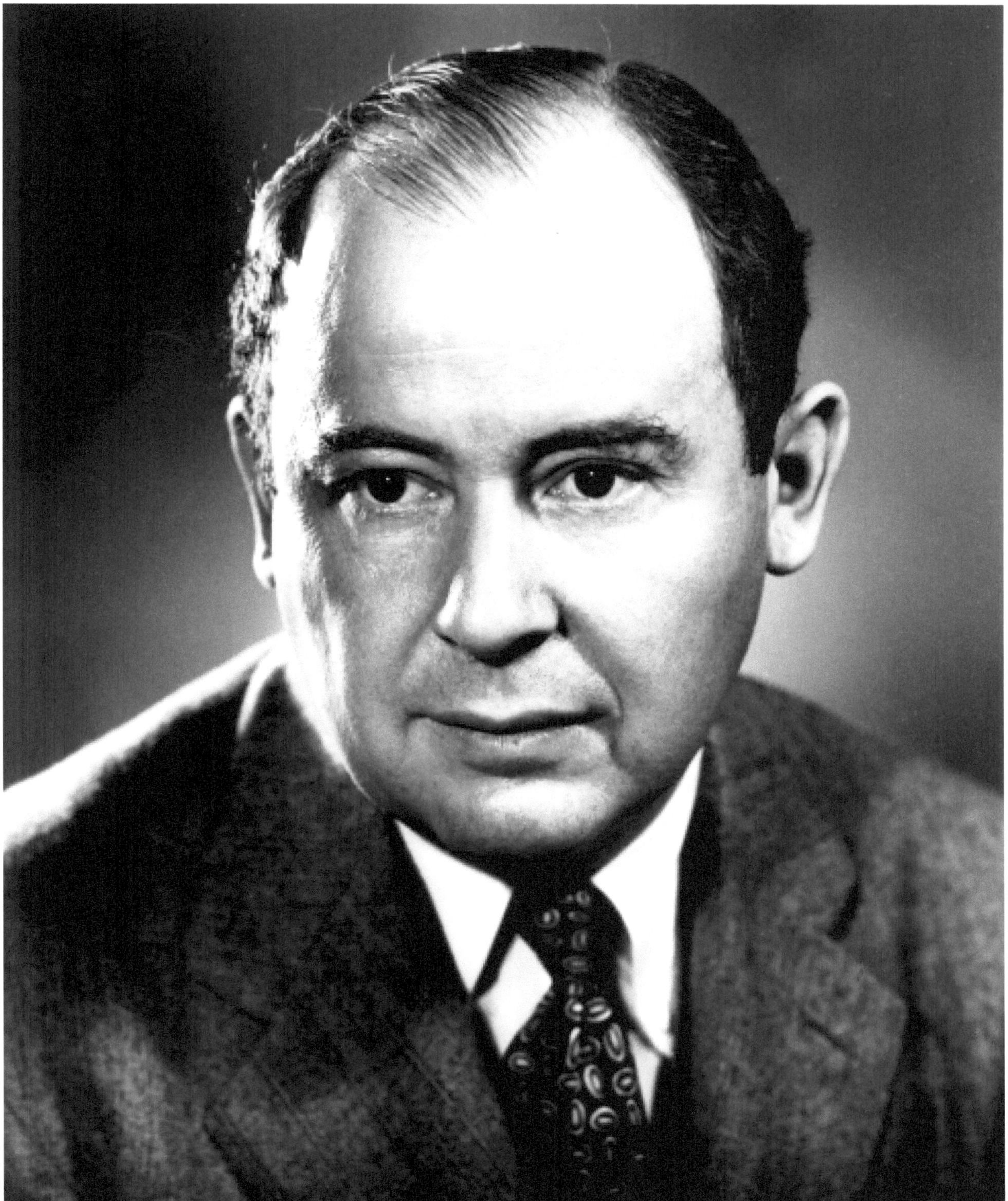

Von Neumann

(1903 – 1957)

Appendix I

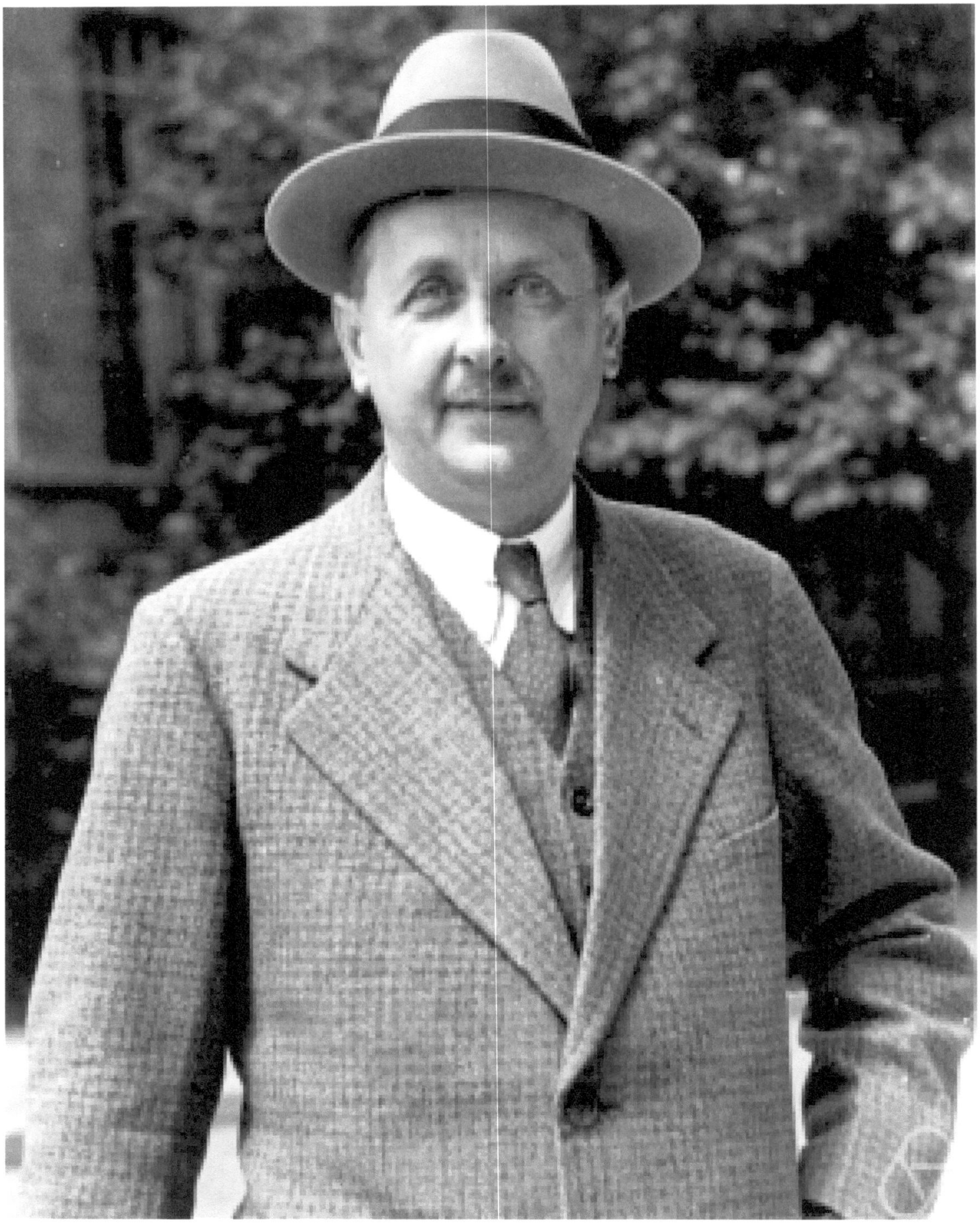

WEYL

(1885 – 1955)

Appendix J: What is the Fields Medal?

The Fields Medal is a major award in mathematics. It is awarded to up to four mathematicians under the age of 40 every four years at the Congress of the International Mathematical Union. There is no Nobel Prize for mathematics, but it is said that the Fields Medal is as close as it gets. The medal is named after a Canadian mathematician and was first awarded in 1936. The obverse side of the medal, seen here, features a depiction of Archimedes and his name written in Greek. The Latin motto can be translated as “To cross his breast and the world possess.” The IMU’s permanent offices are in Berlin, Germany. As of this writing, the last Fields Medals were awarded in 2022. Other prestigious awards in mathematics include the Abel Prize and the Millennium Prize.

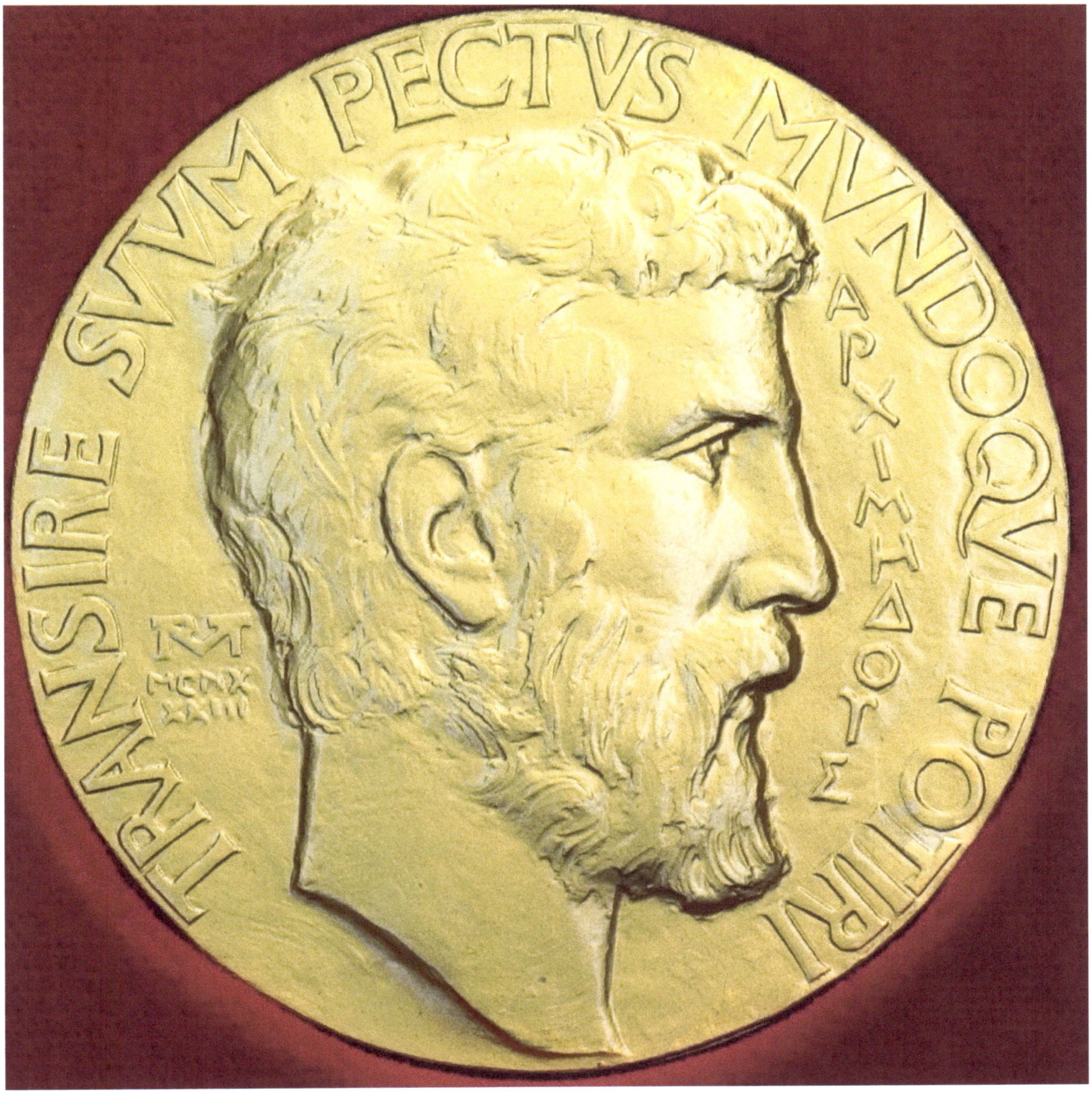

Appendix K: What are complex numbers?

Complex numbers consist of a real part and at least one imaginary part. A simple example is $a + bi$, where a and b are real numbers and i is the so-called imaginary number. In this case, a is the real part, and bi the imaginary part. To enable an algebra of complex numbers, we set i equal to the square root of -1 ($i^2 = -1$). We can make a graph:

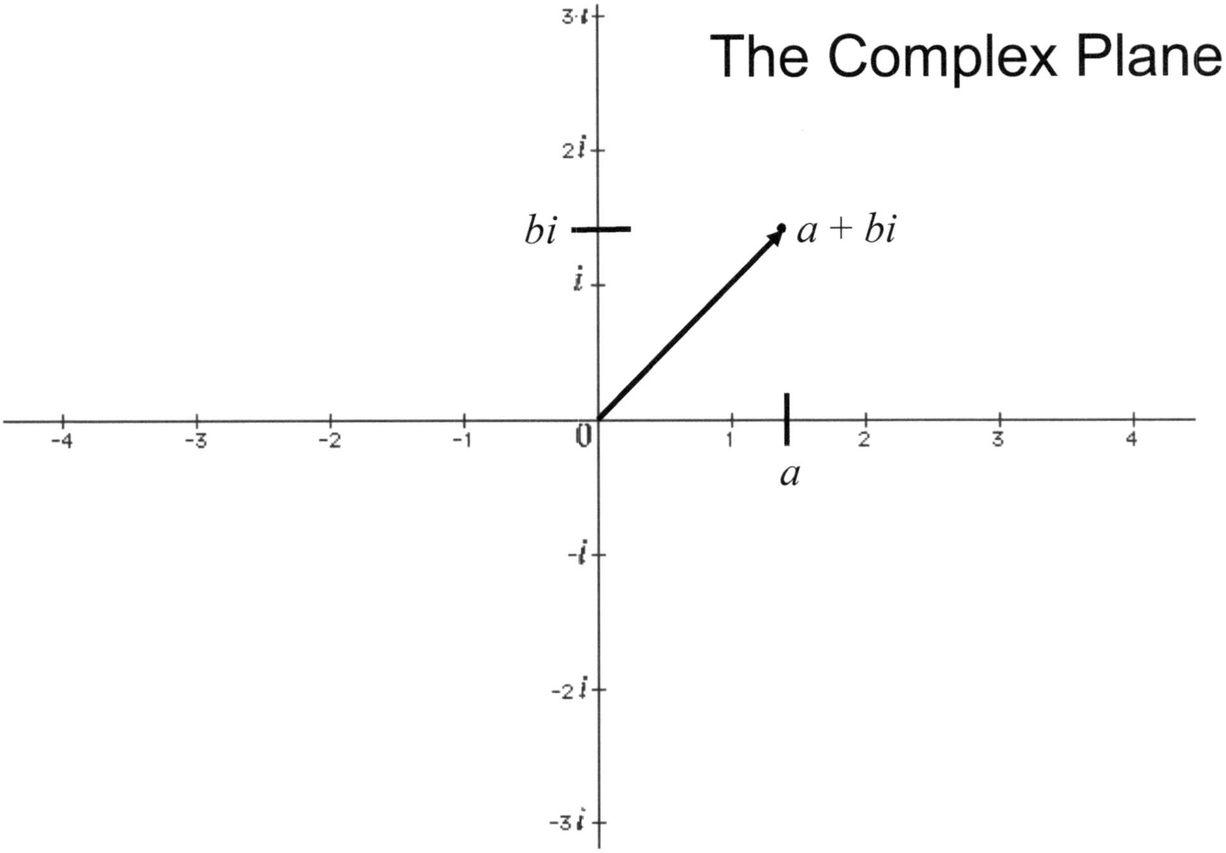

This is the complex plane, not the x-y plane. On it we can plot complex numbers. The horizontal axis is the "real" axis, and the vertical axis, measured in multiples of i, is the "imaginary" axis. The complex plane represents two-dimensional numbers. There are also four-dimensional complex numbers called "quaternions." For quaternions, which are of the form $a + bi + cj + dk$, we have $ijk = i^2 = j^2 = k^2 = -1$.

There are also complex numbers of even higher dimension, such as the eight-dimensional octonions. Are there three-dimensional complex numbers? No, and there are several reasons why not. The main reason is that three-dimensional complex numbers, which would be of the form $a + bi + cj$, cannot form a coherent algebra. In short, there is nothing for ij to equal. Personally, I think this represents the interior of a solid three-dimensional object, but this is only speculation at this point.

Appendix L: What is a function?

In mathematics, a function is an operation denoted by some kind of formula that takes an input and returns exactly one output. When a function *f* acts on an input *x*, we write $f(x)$, pronounced "f of x" (a function of *x*). Functions come in innumerable varieties and forms. If the function squares the input, we write $f(x) = x^2$. Perhaps the function is $f(x) = 867x + 5$. Maybe it's $f(x) = \sin(x/2)$. If we set $y = f(x)$, many functions can be represented on the Cartesian *x*-*y* plane. The function takes an input *x* and returns a *y* value.

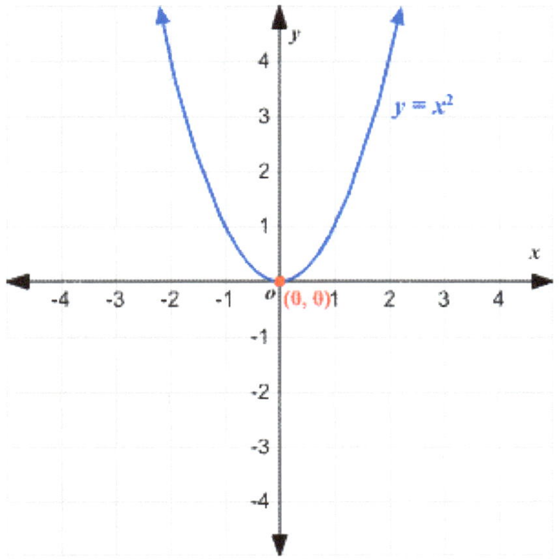

The graph of $y = f(x) = x^2$
$f(1) = 1, f(2) = 4, f(3) = 9$, etc.

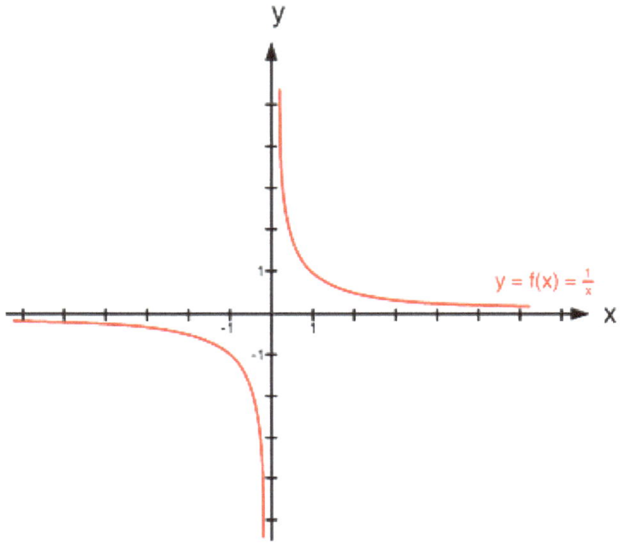

The graph of $y = f(x) = 1/x$
$f(1) = 1, f(2) = ½, f(3) = 1/3$, etc.

There are also functions of multiple variables. In particular, functions of the form $f(x, y) = z$ are three-dimensional, and are plotted on three perpendicular axes: the *x*, the *y* and the *z*. There are also functions of complex variables that are plotted on the complex plane. This is all a simplified description of what a function is, but I am trying to introduce the idea to people who have never heard of it.

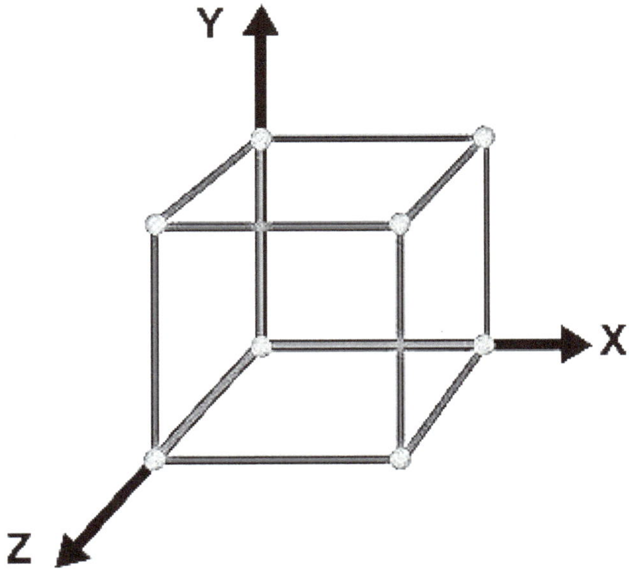

The *x*, *y* and *z* axes

Appendix M: The James Webb Space Telescope

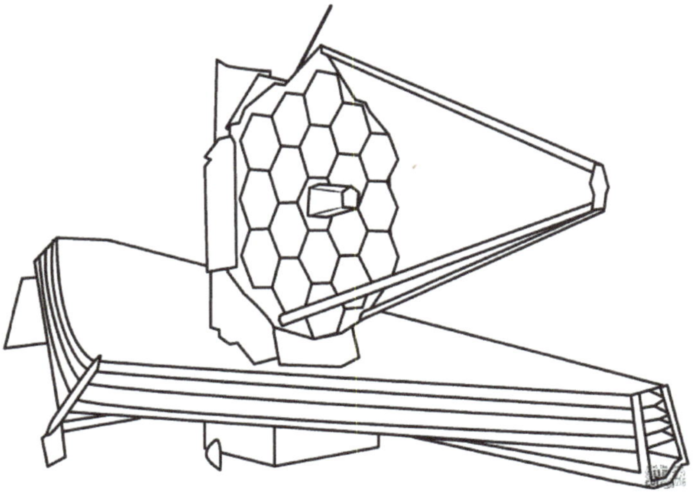

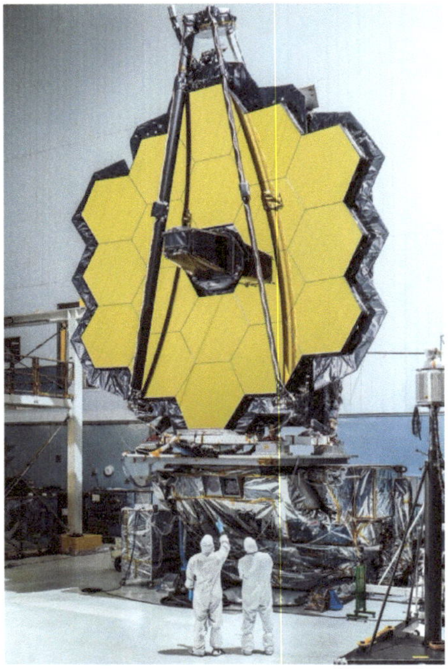

The JWST, launched in 2021 as a joint enterprise of NASA, the European Space Agency, and the Canadian Space Agency, is an infrared telescope designed to look at the dimmest objects in the darkest parts of the sky. It will also be able to examine exoplanets in great detail. It began to send back data from its orbit around the sun in 2022. Its observations of the most distant objects ever seen are causing a bit of an identity crisis in modern cosmology. Nothing the JWST sees at the edge of the observable universe makes sense. I'm not sure we're seeing what we think we're seeing. Then again, DON'T PANIC! James E. Webb was the administrator of NASA from 1961 to 1968. He also held other high-level positions in the US government. He died of a heart attack in Washington, DC, in 1992.

Appendix N: William Shockley

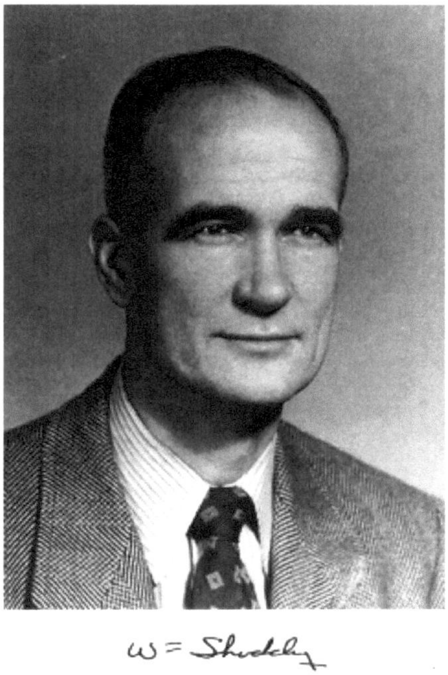

I don't really want to bring this character up, but conscience compels me. History should not be changed. **William Bradford (Bill) Shockley Jr.** (February 13, 1910, London, England – August 12, 1989, Stanford, California) was an American inventor, physicist, and eugenicist. He was one of the most important scientists of the 20th century. He was the head of the research group at Bell Labs in New Jersey that invented the transistor. The word "transistor" comes from "trans-resistor," referring to a voltage applied across a resistor. The transistor basically underlies all modern computing technology. Shockley's team included John Bardeen and Walter Brattain. These three were jointly awarded the 1956 Nobel Prize in Physics for their research on semiconductors and their discovery of the transistor effect. Later in his life, while a professor of electrical engineering at Stanford University and afterward, Bill Shockley became an open and outspoken racist and eugenicist. In 1956, Shockley founded Shockley Semiconductor Laboratory in the city of Mountain View in the San Francisco Bay Area, effectively founding Silicon Valley. Despite his scientific achievements, William Shockley is a figure reviled throughout the world for his avowed racism. He died alienated from his family. These are the facts of the case.

Appendix O: The Sun

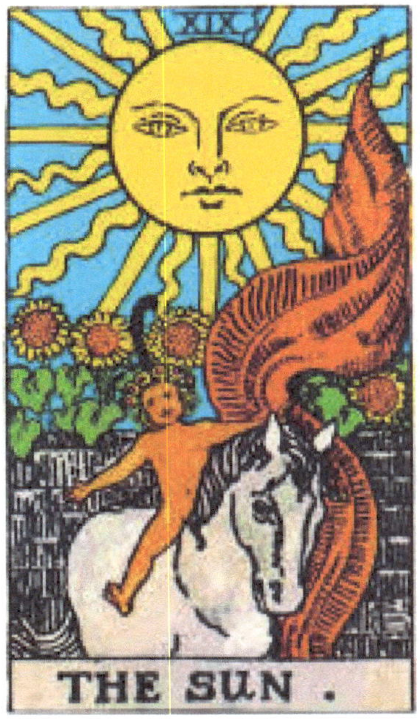

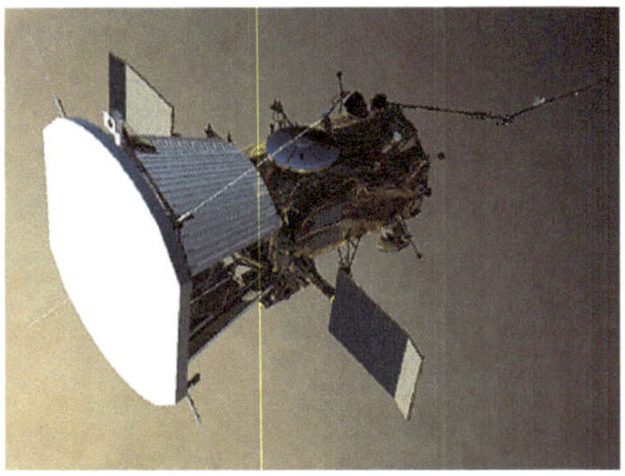

The Parker Solar Probe

"The sun is a mass
Of incandescent gas,
A gigantic nuclear furnace,
Where hydrogen is built into helium
At a temperature of millions of degrees."
-They Might Be Giants

This is not exactly true, as the sun is made not of gas but of plasma, an entirely different state of matter. TMBG wrote a song acknowledging their mistake. The Parker Solar Probe was launched by NASA in 2018. Since then, it has become the closest ever man-made object to the sun and the fastest moving man-made object ever, reaching 430,000 mph (191 km/s) relative to the sun. As of February 2025, it remains in operation and is sending back the closest, most detailed data and images of the sun ever obtained.

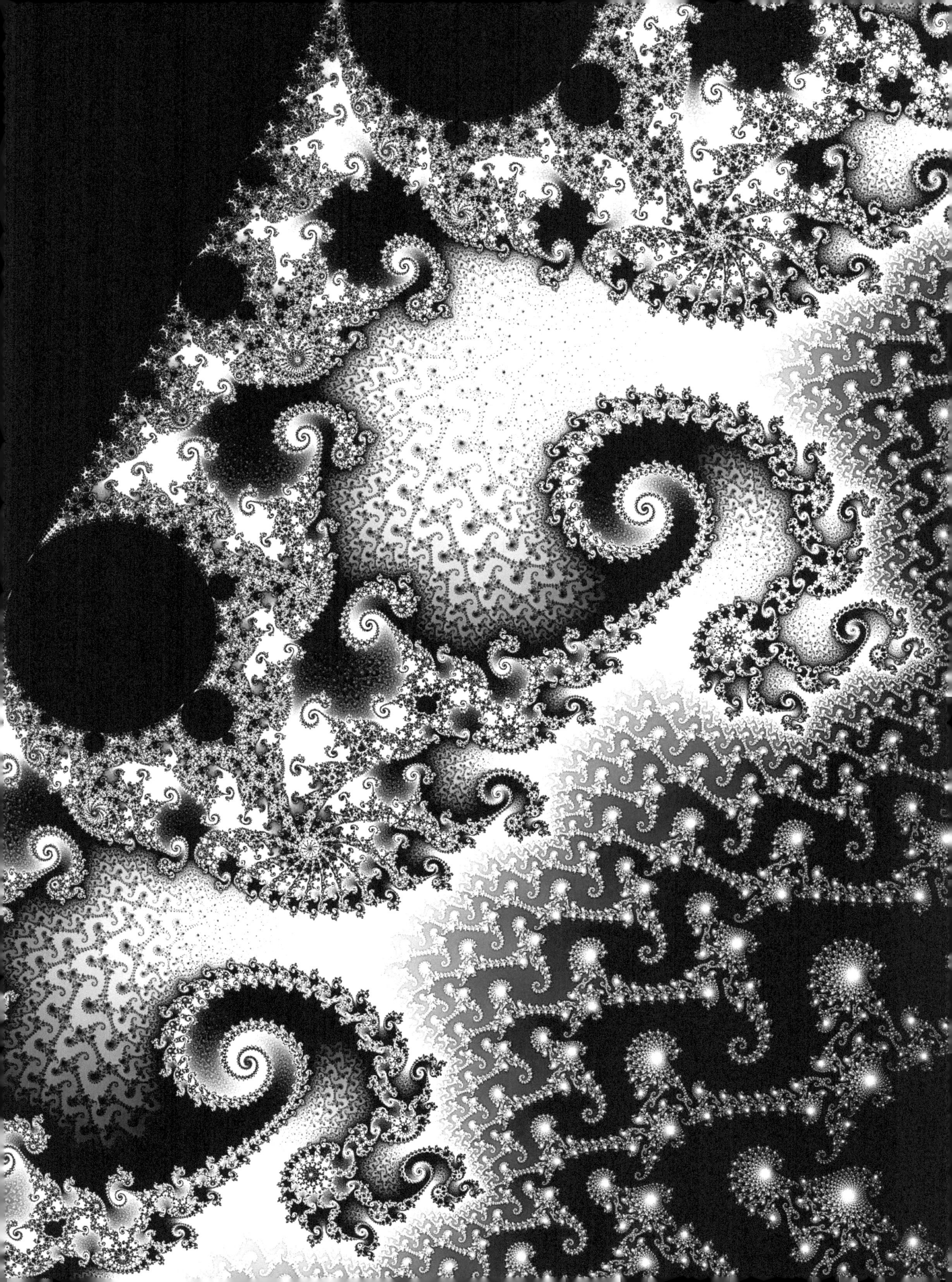

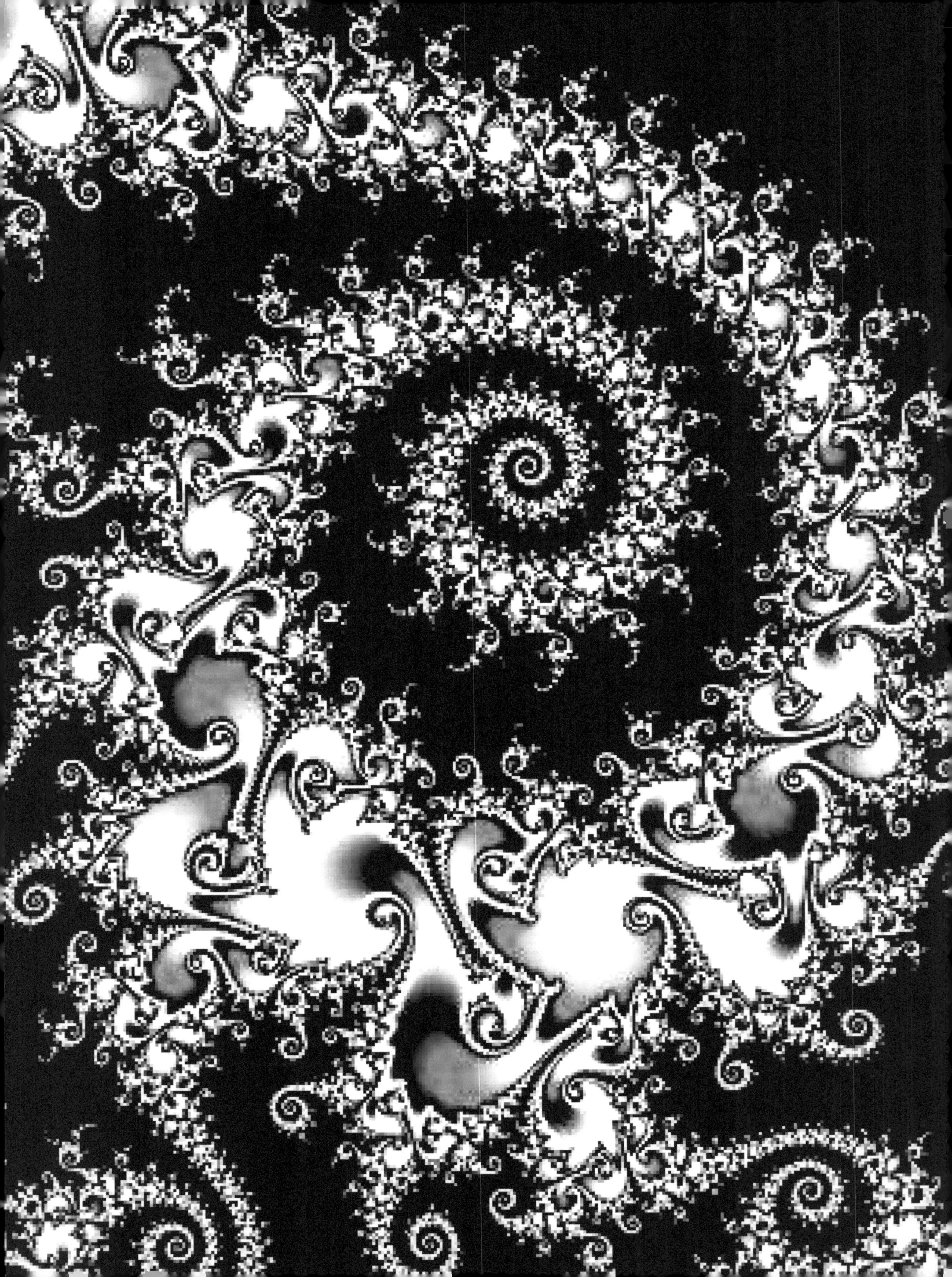

About the Author

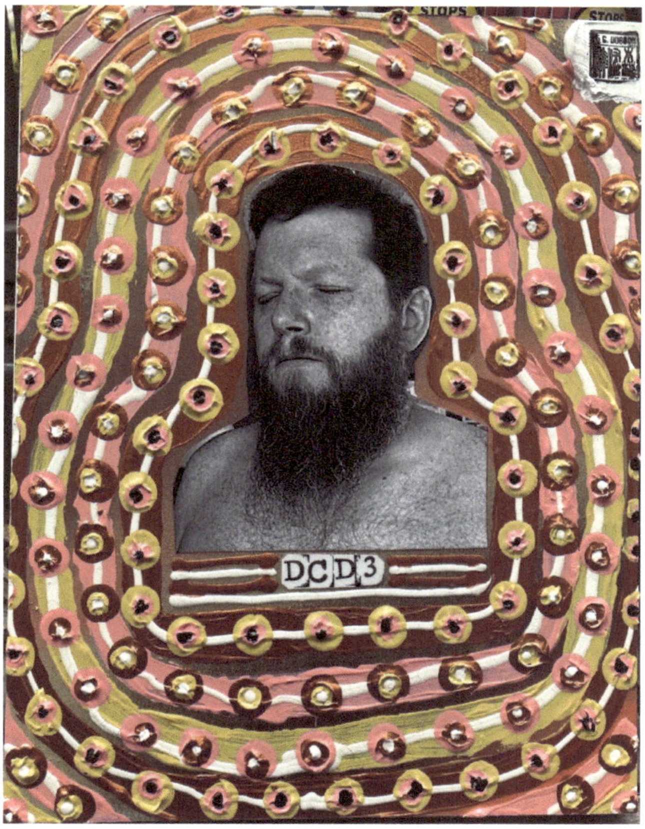

Duncan Calvin Dobson, III, was born in Kuala Lumpur, Malaysia, but has called the San Francisco Bay Area home since the age of three. Duncan got his BA in English with an emphasis in Creative Writing from San Francisco State University in 2008. He also graduated from the California Culinary Academy, which no longer exists.

About the Editor

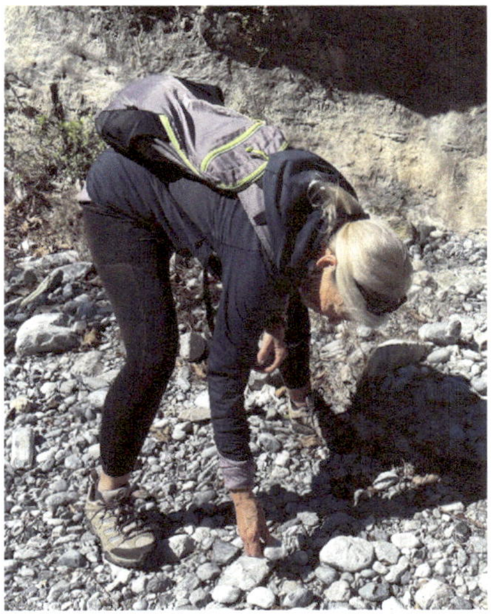

Lucy Dobson was born and raised in St. Louis, Missouri, and has resided in the Bay Area for many years. She travels the world collecting rocks, as seen here on Mt. Olympus in Greece in 2024.

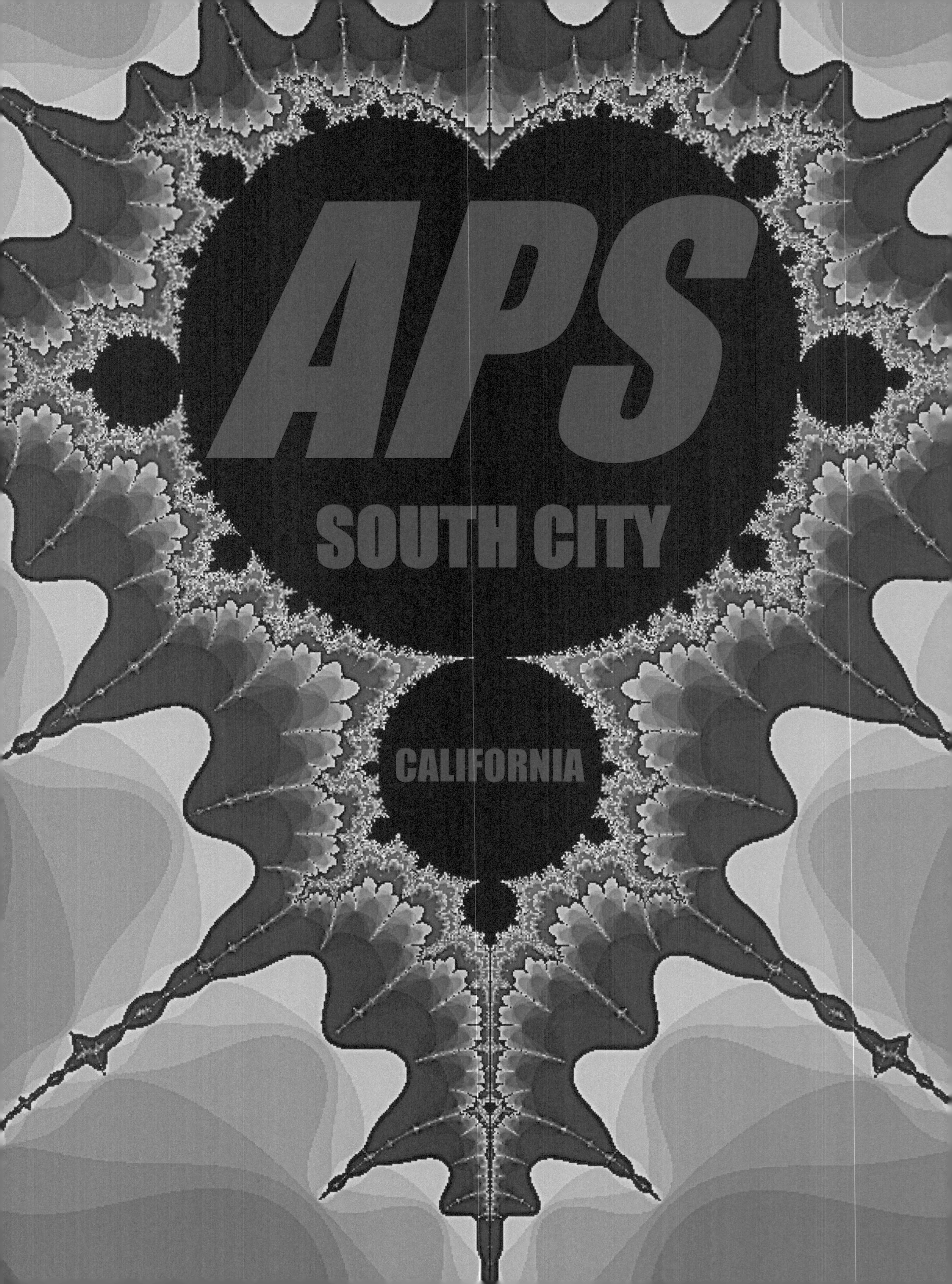
APS
SOUTH CITY
CALIFORNIA

www.ingramcontent.com/pod-product-compliance
Ingram Content Group UK Ltd.
Pitfield, Milton Keynes, MK11 3LW, UK
UKHW061542051026
15625UKWH00011B/584
* 9 7 9 8 2 1 8 4 8 2 2 3 7 *